A NEW APPROACH TO CONSERVATION

For Coco, Mac, Digby, Craggan and Bill,
my canine companions past and present

A New Approach to Conservation

The Importance of the Individual through Wildlife Rehabilitation

GILL AITKEN
Lancaster University

Routledge
Taylor & Francis Group
LONDON AND NEW YORK

First published 2004 by Ashgate Publishing

Reissued 2018 by Routledge
2 Park Square, Milton Park, Abingdon, Oxon OX14 4RN
711 Third Avenue, New York, NY 10017, USA

Routledge is an imprint of the Taylor & Francis Group, an informa business

First issued in paperback 2018

A Library of Congress record exists under LC control number: 2003058294

ISBN 13: 978-0-815-38739-8 (hbk)
ISBN 13: 978-1-138-61870-1 (pbk)
ISBN 13: 978-1-351-16356-9 (ebk)

Contents

Preface

In 1995, the British government backed a conservation scheme – supported by, amongst others, the Royal Society for the Protection of Birds (RSPB) and the Wildlife and Wetlands Trust – to save the Spanish white-headed duck. This was, ostensibly, a laudable conservation objective. The threat facing the white-headed duck was from hybridisation with the ruddy duck, numbers of which were said to migrate in the spring from Britain to Spain. The proposed solution to this conservation problem was the slaughter of the 3,500 ruddy ducks resident in Britain. This was part, only, of a much larger European initiative involving Spain, Portugal, France and the Netherlands. However, the search for a solution continues. In 1999 the Government undertook a three-year trial to assess the feasibility and cost of ruddy duck elimination, and the results of this study have recently been published. The pertinent features of the ruddy duck case are first, that the number of white-headed ducks (19,000) is low (the species is threatened with extinction); second, that the number of ruddy ducks (500,000 in their native North America) is high (the species is common); and third, that those ruddy ducks targeted for slaughter are alien immigrants to Britain, while the white-headed duck is a native of Spain (Brown, 1995).

It would be unfair to argue that all conservation bodies were (or are) in support of this proposal (or indeed others of its kind). But it does seems fair to suggest that, as it was supported by a number of high profile conservation bodies as well as the government's own Department of Environment, it is a scheme that accords with a widely-recognised and respected conservation rationale. This proposed scheme could be argued to characterise mainstream conservation's primary concerns: rarity, extinction and nativity. These concerns are substantiated in efforts to implement native, rare species protection (thereby heading off extinction), regardless of whether individuals of other species are harmed in the process. All of these concerns are in turn subsumed under the more general concern of protecting biodiversity.

That the concerns identified in the proposed ruddy duck culling programme are typical of conservationists can be illustrated by way of a few examples. The World Wide Fund for Nature (WWF), one of Britain's leading conservation bodies is primarily geared towards 'fighting for our vanishing species' including, for example, the highly endangered Indian tiger, African black rhino and European wolf (WWF promotional leaflet). A number of conservation bodies focus expressly upon aiding particular threatened species, such as the charities 'Butterfly Conservation' or the 'International Otter Survival Fund'. But even where the remit of an organisation is broader – such as that of the RSPB, which is dedicated, amongst other things, to protecting birds' habitats – emphasis is placed upon rare habitats and habitats for rare species.

The RSPB's concern for species and habitats sharply contrasts with their indifference towards the fate of individuals. Similarly, English Nature, the government-sponsored conservation body, concerns itself with species but not individuals. As the organisers of local bat warden schemes, they finance a help-line for dealing with public bat-related enquiries. This does not cover the cost of dealing with injured or displaced individual bats.

Conservation bodies are typically concerned about – and with – endangered species. They are also typically interested in saving only those species that are native. To give just one example, conservationists in Britain are making strenuous efforts to protect the disappearing native red squirrel, while the thriving grey squirrel, introduced from North America in the late nineteenth century, is heavily persecuted (Evans, 1996). To be fair, while conservation bodies typically make positive efforts to save only native species, most actively pursue non-native species destruction only when there is a conflict between the interloper and a native species or habitat. Such is the case, with the two squirrel species in Britain and with, perhaps the best-known example of all, the rabbit in Australia. But some conservation bodies have a more extreme stance. For example, the Ornithological Union favoured eradicating from Britain not only the ruddy duck, but all introduced species, including the naturalised (and perfectly harmless) snow goose, bar-headed goose, mandarin duck and parakeet (Brown, 1995). And certain eradication or reduction programmes are routinely supported by mainstream conservationists. These include programmes involving the grey squirrel, mink, magpie, ruddy duck, Canada goose, and red deer.

The same ideology is to be found in conservation attitudes towards introduced plant species. Conservationists make strenuous efforts to eradicate or reduce Bracken, Japanese knotweed and *Rhododendron ponticum* (Evans, 1996). The difference, of course, is that with the former but not the latter examples, these actions are carried out in spite of the harm done to individuals capable of suffering. Conservationist's support of culling and eradication programmes is evidence of their willingness to prioritise the species or habitat over the sentient individual. This attitude is also evident in, for example, the RSPB's apparent indifference towards the fate of injured birds: caring for individual birds is not part of their remit. It is also evident in the attitude of English Nature who, as mentioned above, recompense their bat workers for expenses incurred in dealing with bat-related calls, except when these involve orphaned, injured or grounded bats. For English Nature, what is important is the continued survival of each species of bat. What is of greatest concern to them is biodiversity.

Despite the differences between the specific remits of conservation bodies, there is much evidence to suggest that their fundamental concerns are those relating to species – species rarity, extinction and nativity. These are concerns precisely because biodiversity is important to them. It is not merely conservationists on the ground who share these concerns. Such priorities are evident amongst academics too – biologists and philosophers alike. For example, in *Essentials of Conservation Biology*, Richard Primack (1993) suggests there are at least five underlying assumptions that seem to be held by conservationists with little – if any – exception, and which form the basis of conservation biology:

- The diversity of organisms is good;
- The untimely extinction of populations and species is bad;
- Ecological complexity (e.g. co-evolution and ecological relationships) is good;
- Evolution is good;
- Biological diversity has intrinsic value (pp. 19–20).

Philosophical analysis of conservation issues seems, too, to rest upon – or challenge – to a large extent, these very assumptions. For example, Norton (1987) considers the reasons for preserving biological diversity while Ehrlich and Ehrlich (1981) have explored the implications of extinction. J. Baird Callicott's (1987) arguments depend heavily upon the idea that ecological relationships are a conservation value, and he discusses the intrinsic value of ecosystems and species in, for example, *Environmental Ethics* (1980) and *The Preservation of Species* (1986).

Conservationists seem to accept, quite readily, widely held assumptions about the value of both biological diversity and, primarily, (native) species. They accept, too, the assumed need to address extinction and rarity. Yet it is the role of environmental philosophers to challenge such assumptions. Sometimes this challenge is partially met, as evidenced by the title of Norton's (1987) book *Why Preserve Natural Variety?* But such critical appraisals rarely go far enough. They tend to consider why certain values are important to conservation *as it is currently recognised.* They rarely question whether conservation might be differently understood. They begin, typically, with a particular idea of conservation in mind and then seek to justify that idea or model. They seem, too, to miss interesting questions by stopping short in their enquiries. For example, in examining the assumption that 'the untimely extinction of populations and species is bad' the question of how, if at all, human-caused extinction is to be differently appraised from other extinctions is ignored. Similarly, though questions about, for example, the value of biological diversity have attracted some investigation, the related issues of rarity and nativity are seldom addressed. Yet these are the very topics that are regularly featured in conservation policy and management plans. And although there have been many discussions about the conservation value of species, there is rarely any discussion about whether the *category* of species is an appropriate focus of value.

On the basis of these observations, it seems possible to identify three assumptions that are shaping conservation's theory and practice:

1. That holistic units – in particular species – are conservation's units of importance.
2. That certain categories of such units – namely the rare and the native – are of particular importance, and that their opposite numbers – namely the common and the introduced – are of no (or even negative) conservation value.
3. That non-human individuals have no conservation value.

The task of this book is to examine the validity of these assumptions. This I will do through analysis of the concepts (e.g. rarity) themselves, but also through a careful study of the practice of wildlife rehabilitation. Though dealing with some of the 'players' in the conservation arena – that is, with wild animals – wildlife rehabilitation is a practice that is said by many to have little or no (perhaps even negative) conservation value, and this primarily, because it is concerned, not with species, but with individuals. It is also because it accords no special value to the rare or the native. The set of values underlying wildlife rehabilitation are very different from those underlying conservation – but I choose them with purpose. They serve as a way of criticising the values implicit in the conservation rationale as it currently stands. They are also good examples of values that, I will go on to argue, are more coherent and more appropriate for a conservation rationale.

In criticising some of the values assumed by conservationists – and thus in questioning the current conservation rationale – I am clearly saying something about the conservationist's understanding of what it means to conserve nature. I am commenting on conservation's fundamental concern – the desire to protect biodiversity. My critique of the conservation rationale is not intended to dismiss the value of biodiversity, to claim that it ought to no longer be a primary concern of the conservationist. Rather, in revealing conservation's underlying assumptions as problematic, I demonstrate that the concept of 'biodiversity' has to be differently understood. I do not reject the notion of 'biodiversity' out of hand as either meaningless or unimportant, but I do argue that it's meaning does need to be modified. Currently, it is too narrow, too limited in scope.

This book, which is divided into four parts, begins, in Part One, Chapter One, with an implicit acknowledgement of the value of biological diversity. That is, I begin by accepting that if we wish to safeguard nature as we know it, we must safeguard its diversity. However a challenge is immediately offered in the form of a question about how, if at all, human-caused extinction is to be differently appraised from other extinctions. If it cannot be, there may be problems in arguing that conservation is necessary in order to safeguard nature. I find, however, that because of the nature and effects of human-caused extinction, intervention of some kind is essential if we wish to safeguard the natural world. So my starting point is with familiar territory: conservation action is essential if we are to retain nature in a form we currently recognise. An integral element of this is its diversity. From here on I part company with familiar rhetoric and assumptions. In Chapters One and Three I examine the concepts of 'rarity' and 'nativity' respectively, in order to determine their proper place in the conservation rationale. It turns out that rarity does not deserve the attention it currently enjoys and that there is much more room for valuing the common or the abundant than usually supposed. I find, too, that the current understanding of the term native is too rigidly construed. A more inclusive understanding of nativity – focusing on the idea of belonging – is both more coherent as well as useful in directing conservation action. It is an understanding that makes room for species that, though introduced, have become familiar and valued.

In Part Two I shift emphasis, taking time to explore both the role and primary goal of the conservationist. My aim is to show that some of the values underlying the current conservation rationale ought to be replaced by different values – in effect, that the conservation rationale requires revamping. But before I can argue about the substance of the conservation rationale, I need to be quite sure about the kind of role we ought to have in effecting conservation. The purpose of Chapter Four is to establish the nature of both the theoretical as well as the practical relationship between humans and nature. In Chapter One, I show that extinctions brought about by humans are different from other extinctions – a claim that assumes humans act upon nature and are not merely an aspect of nature. This idea is further supported by the finding that humans and nature are conceptually separate. And my claim, from Chapter One, that we need to intervene in order to remedy the current situation is shown, in Chapter Four, to be a sound one. Intervention is the only coherent conservation position.

The fifth chapter deals with the important question of how to best safeguard nature. I suggest that 'wildness' is the key factor. By safeguarding wildness, I argue, we can best safeguard the nature we currently cherish. 'Wildness' encapsulates the spirit and meaning of nature in a way that 'biodiversity' does not. And the aim of protecting and promoting wildness better fits with the process by which (again, argued in Chapter One) nature is best safeguarded – that is, with speciation.

Part Three of the book returns to my critique of the conservation rationale. Chapter Six examines the assumption that 'species' denotes the unit of importance in conservation. I find that the category of 'species' is much less concrete than supposed. This being so, I argue that it would be reasonable to place less emphasis on species boundaries than conservationists typically do. A much more fruitful focus for the conservationist must be upon the opportunity for speciation. It is the perpetuation of difference that safeguards biodiversity – that is, the ongoing process of differentiation – rather than a focus on boundaries that can never be truly defined. I detect a stark contrast between the contrived nature of the category 'species' and the concrete nature of the category 'individual'. This leads me to suggest that the latter category deserves greater attention than it currently enjoys. Indeed, in Chapter Six I go on to articulate reasons why it is in the interests of conservationists to value (sentient) individuals a great deal more than they do. I carry the theme of valuing the individual further in Chapter Seven. I argue that the practice of wildlife rehabilitation – a practice that deals with some of the 'players' in the conservation arena (wild animals), but as individuals and not as species – has a great deal more value to conservation than conservationists generally believe.

My proposed new emphasis upon the individual in conservation is put to the test in Chapter Eight where potential difficulties are identified and addressed. I deal with and dismiss worries concerning first the impact of wildlife rehabilitation upon the evolutionary process, and second, the conflict between rehabilitation's care for the individual and conservation's culling of individuals. The chapter ends with a case study – oiled bird rehabilitation. I demonstrate that even with 'hard cases', rehabilitation's principle of individual care is a great deal more compatible with conservation's objectives than critics are willing to acknowledge.

In the final section of the book I begin to pull the various threads together. I earlier argued both that conservation is best achieved by the protection and/or promotion of wildness and that wildlife rehabilitation has conservation value. In Chapter Nine I address the potential tension between the call for wildness and the intensive management characteristic of wildlife rehabilitation practice. I gain both a detailed understanding of what we mean by wildness, and an insight into the effects of human contact upon rehabilitated wild animals. It emerges that the practice of wildlife rehabilitation is perfectly compatible with the conservation remit of promoting wildness.

The tenth and final chapter considers the implication, for the conservationist, of what we might call 'the new conservation'. Here, the arguments from the previous chapters are pulled together to form a coherent conservation strategy for the future. This is a strategy that no longer focuses so uncompromisingly upon 'species' or upon 'the native' and 'the rare'. It increasingly focuses, instead, upon the value of the individual, the common and the introduced. These values are not only compatible with, but also help to *define* the underlying conservation value of biological diversity.

I arrive, ultimately, at a somewhat different understanding of nature conservation than we have at present. It is a much wider, more inclusive understanding. Although the conservationist's current fundamental value – biodiversity – has been retained as a conservation value, I reveal that it is too narrowly construed. There are more coherent ways of respecting and maintaining biodiversity – and thus of safeguarding nature – than the conservation rationale presently allows for. There should be more room for the importance of the individual, the commonplace and the introduced. There should also be more room for the importance of relationship, and for ecological processes such as speciation. Nature conservation, I argue, ought to be pursued in a quite different manner than is currently the case.

Acknowledgements

Numerous people have played a part in helping me form my ideas for this book and I thank them all. I particularly wish to thank Alan Holland and Kate Rawles for many stimulating discussions and helpful feedback over a number of years, and Steve Powell for broadening my understanding of the issues surrounding wildlife rehabilitation. I am especially indebted to Alan Holland who has always been a great inspiration to me. Without his unfailing encouragement, enthusiasm and wisdom I would certainly have fallen by the wayside.

I also wish to thank the British Academy for helping to fund the original thesis, on which this book is based, and the Arts and Humanities Research Board for funding me while I was preparing the book for publication.

Chapter One was first published as 'Extinction' in *Biology and Philosophy* (Vol. 13, No. 3, 1998) by Kluwer Press. Part of Chapter Eight was first published in *Environmental Values* (6, 1997) as 'Conservation and Individual Worth' by Whitehorse Press.

PART I
EXPLORING CONSERVATION'S BASIC ASSUMPTIONS

Chapter 1
Extinction

Introduction

The extinction of species is of fundamental concern to conservation bodies. For example, The International Union for Conservation of Nature and Nature Resources (IUCN) keeps detailed information on the status of species threatened with extinction in their Red Data Book series. Analyses of the data 'stimulate conservation activities, the results of which are then fed back into the system to determine new conservation priorities' (IUCN, 1988).

There is a clear message here: certain forms of species extinction are undesirable. The belief – which is implicit in the work of all conservation bodies (it being a corollary of the conservation aim to safeguard biological communities) – is infrequently questioned or rarely justified. When justifications are offered, anthropogenic extinction is usually assumed to be significantly different in kind from extinction caused by other means. This is seen as one reason to condemn it. (Condemnation does not, of course, *necessarily* follow even if significant differences are found; this would require further argument.) The assumption that the two kinds of extinction differ is even less likely to be argued for. And where arguments are found, they are usually unsatisfactory and often appear confused. If good arguments in support of this assumption can be found, the conservation rationale of working to prevent the extinction of species – and therefore of safeguarding biological communities – would seem to be strengthened. In other words, this would lend support to the conservation rationale as it currently stands. What I intend to do in this chapter is consider whether – and in what ways – human-caused extinction may be different from other forms of extinction. I can then decide what impact this may have upon ideas about conservation, and in particular about saving species.

Placing anthropogenic extinction

Before I can tackle the question of why current species extinction is thought to be undesirable, I need to address the more fundamental question of whether extinction caused by humans really differs in kind from other forms of extinction. The other forms of extinction fall into two camps: background extinction – a low-grade continuous form that includes the modification of one species into another as well as the replacement of species with better-adapter competitors – and mass extinction – a rapid, wide-ranging and catastrophic form that takes no account of a species' adaptations.

My first line of approach is to determine whether the current anthropogenic extinction event can be aligned with either of these other forms of extinction in terms of scale or rapidity. In order to do so, I need to make use of the available data on extinction rates. This presents a potential problem: there are substantial discrepancies in the data. Some of these discrepancies are merely apparent. Since there seems to be no standard way to describe extinction rate, one source might base its figures solely on mammal extinction, while another might consider the extinction of mammals and birds. There are, however, real discrepancies that arise as a result of our limited knowledge in this area and the need for estimates. Nevertheless, despite the often quite substantial discrepancies, we know enough to say that, in terms of rate, the current anthropogenic extinction event is sufficiently dissimilar to background extinction to justify a different judgement about its significance. Its rapidity, though, is sufficiently close to that of past mass extinction events to justify a similar judgement.

Primack (1993), for instance, estimates (on the assumption that there are around 10 million species on Earth today) that we would expect to lose between one and ten species a year through natural background extinction. But he notes that the current observed rate (based on birds and mammals) is 100 to 1000 times greater. Not only is the current rate of extinction many times greater than background extinction, it is also increasing. The United Nations Environment Programme's estimate of 1995 confirms that '[n]early three times as many birds and mammals (112 in all) became extinct between 1810 and today than were lost between 1600 and 1810 (38 species)' (Porritt, 1996). This represents 'an extinction rate of 50 to 100 times the expected natural rate'. While rather more conservative than Primack's it is still surprisingly high. According to a 1992 estimate of the rate of (total) species extinction from the World Conservation Monitoring Centre (Groombridge, 1992), one hundred thousand species are lost a year. This works out as 273 species a day or 11 species an hour.

Current anthropogenic extinction is of a quite different order to background extinction. Even on a conservative estimate it is on a much greater scale. It fits more easily into the mould of 'mass extinction', such as occurred at the end of the Permian period (250 million years ago) when, as Primack tells us, '50% of animal families, including over 95% of marine species' disappeared. Although these percentages are much higher than current losses, numbers of species lost, according to Norton (1986), are thought to be similar. This is presumably because of today's greatly increased diversity. So if the current extinction event resembles past mass extinction events in terms of scale and rapidity, these features cannot be upheld as evidence of a difference between them. Anthropogenic extinction conforms, in these particulars, to a familiar pattern of extinction. If rate of extinction is unhelpful as a distinguishing feature, might there, then, be other factors that would indicate a difference between the current extinction event brought about by humans and past mass extinction events?

Are anthropogenic extinctions unnatural?

It would not do to talk about possible differences between forms of extinction and neglect to mention the distinction between natural and unnatural occurrences. This is an area covered by authors as diverse as Elliot (1994) and Evernden (1985). In the present context, Wilcox (1988) notes that '[t]he argument is frequently made that extinction is natural'. The point of such an observation is that if extinction is natural, and if we value what is natural (as conservationists appear to do) we ought to find extinction, if not *valuable*, then at least acceptable. And within the idea that natural processes ought to be acceptable there lurks a feeling of inevitability: that to try to resist the outcome of such events would be quite pointless.

The argument that we ought to accept extinction if we value what is natural is, at best, unhelpful. The belief that some extinction is natural does not commit us to a belief that *all* extinction is necessarily natural. Indeed, the reason that is usually given for the former belief is that extinction occurs in the absence of human agency.[1] It would be nonsense to argue that anthropogenic extinction is natural on the grounds that non-anthropogenic extinction is natural. And if anthropogenic extinction is natural, it would not necessarily follow that it should be found acceptable. There is no necessary connection between the two. And if events that occur as a result of human agency are natural, it is difficult to see what might be excluded from the natural category. The observation that anthropogenic extinction is natural would be claiming nothing of interest.

It may well prove worthwhile to pursue the 'natural' question in a clear-headed way, with respect to seeking a difference between past and present extinction events. However, I am going to take a somewhat different approach. Whether the current anthropogenic extinction event is natural or not, the question I am going to ask here is whether it is *unprecedented.*

The significance of extinction rates

One stance – and that taken by those we may call the traditionalists – is to argue that whilst the current extinction rate is clearly too rapid to belong to background extinction, its very great rapidity also puts it beyond the realms of past mass extinctions; it is of a greater order of magnitude again. Primack (1993) writes that 'the current rate of extinction of species is greater now than at any time in the past'. And Norton (1986, p. 270), talking of past mass extinctions, is convinced that '[w]hile these events may have approached currently expected species losses in total numbers, they were spread over much longer periods of time, measured in millennia'.

Most biologists seem to adhere to this model for the majority of past mass extinctions. However, there is now an alternative theory for (at least) one of the extinction events: the mass extinction marking the end of the Cretaceous period 65

---

1 There are, in fact, two separate stances: that humans, belonging as they do to culture, cannot be said to be natural beings, and that human agency is unnatural.

million years ago, when the dinosaurs (and other less well known species) disappeared. This alternative theory suggests a much more rapid demise for the dinosaurs. Their demise is said to result from a 'nuclear winter' brought about by an asteroid or comet hitting the earth.

Opinion is divided amongst scientists. Many earth scientists, according to Bowler (1992), dispute the evidence for such impacts. Other scientists, as in the case of Campbell (1987), accept the occurrence of impacts but claim that such an impact would merely have added the finishing touches to an extinction that was already in full swing. Yet other scientists seem to accept the theory. Primack (1993, p. 79), for instance, says that a collision with an asteroid resulting in climate change and extinction was quite likely. And Gould (1989, p. 305) believes that 'mass extinctions are *more frequent, rapid, devastating in magnitude, and distinctively different in effect* than we formerly imagined'. The crucial point is that if the theory is correct, dinosaur extinction 'may have taken less than a year' (Begon *et al*, 1990, p. 842). And if this is so, there is at least one mass extinction event of the past that may not be able to be distinguished in effect from current human-caused extinction in terms of rapidity. It may even have occurred more quickly.

It would be valid for me to point out here that current rates of extinction are almost certainly vastly underrated, and this is because many species are as yet unknown and unrecorded. Estimates of the number of species currently in existence range from five million in total (Primack, 1993), to 30 million species of insects alone in tropical moist forests (Wilcox, 1988). Wilcox notes (p. ix) that '[w]hen the tropical moist forest tallies start coming in, the documented extinctions will undoubtedly take another order of magnitude jump over the historic rate' but that for the moment, 'for every species listed as endangered or extinct at least a hundred more will probably disappear unrecorded'.

Nevertheless, it is generally accepted that there are gaps in the fossil record, that some past extinctions remain unrecorded too. Since our current level of knowledge is so very sketchy about past and present species numbers, and since we are unable to do more that guess about extinction rates (factors which must be at the root of the discrepancies in the literature) then it is not possible to base any conclusions about similarities or differences between past and present mass extinctions upon extinction *rates*. We must look elsewhere.

In what follows I will argue that there are significant differences between the present anthropogenic mass extinction event and past mass extinction events that are unrelated to quantity or rate. I identify four such differences. The differences relate to kinds, persistence, extinction models, and modes of impact. Their significance lies in the fact that, together, they lead me to question the validity of extrapolating from past mass extinctions to the present extinction event. In particular, they call into question the validity of predicting a similar post-extinction recovery.

Grounds of distinction

A difference in kind

Current anthropogenic extinctions and past mass extinctions are, I propose, different in *kind*. The grounds for affirming such a difference are as follows:

- Suppose we take the asteroid theory – as an explanation for the dinosaur extinction – to be the right one. Given that we have not ruled out the possibility of anthropogenic extinctions as natural (or even assuming that they are) I can observe that humans and asteroids are quite different kinds of natural cause. The latter, though clearly natural in that they are 'of the universe' are nevertheless alien, extraterrestrial and non-living. Humans, on the other hand, are thought of as natural (at least in part) in that they are familiar, terrestrial and living. Asteroids do not belong to the subset of 'the natural' that constitutes the system of our natural world that is undergoing evolutionary change. In that respect they are different, even, from the (extraterrestrial) sun whose input is, of course, integral to the system.
- The thrust of the above argument can be used in a weaker form to argue for differences between extinction events caused by humans and past mass extinctions excluding those brought about by asteroids. The major causes of past mass extinctions – plate-tectonics, climate change (and to a lesser extent volcanic eruptions) – although neither alien nor extraterrestrial are clearly non-living. They are, of course, a part of our natural world in one sense, but they belong to a category that is separate and distinct from the category to which humans belong: while humans are organic, these other causes of extinction are inorganic. They have not, as humans and other organic things have, evolved through a process of natural selection. And, although connected to the process of evolution by way of cause and effect, these inorganic factors are not bound up with the evolutionary process in the same way as are organic ones. Indeed, in that they constitute *events*, the inorganic factors act upon and are moulded by evolution, but organic entities are the very stuff of evolution. There is a sense of the organic as internal to the system and the inorganic as external to it. And so anthropogenic extinction, as the only organic cause of extinction, is quite distinct – in being internal to the system – from the other, inorganic, external causes of extinction.[2] It is further set apart in being the only extinction event that has been brought about, not only by a species, but also by a single species.
- Assuming that a meaningful distinction can be made between that which is internal and that which is external to the system, I note a further difference in kind. This relates to the different natures of the forces. Humans, as internal to the system, denote a single species gradually expanding its niche, colonising all

---

2 This distinction may be challenged by, for example, proponents of the Gaia hypothesis who maintain that there is a two-way evolutionary interaction between the organic and inorganic (with, presumably, a loss of differentiation between that which is internal and that which is external to the system).

habitats, driving out other species and eventually engulfing all other niches. This is quite different in nature to the external force that rips through habitats, devastates them and then is gone. The difference may be thought of as analogous to external injury and internal disease.[3]

Given that anthropogenic extinctions are different in kind from past mass extinctions, there are grounds to question the validity of extrapolating from past extinction events to the present extinction event. The argument would not be that different causes necessarily guarantee different effects, but rather that given different causes it would not be safe to infer the same effects. The outcome of the current extinction event cannot be assumed to resemble that of past mass extinction events.

Persistence is unprecedented

Anthropogenic mass extinctions differ – *or will have done if we don't change our ways* – from past mass extinctions in their *persistence*, a persistence that is unprecedented. For as long as human populations increase (or even remain constant) and we continue to behave as we do, we will without doubt continue to destroy and fragment other species' habitats for our own use. Consequently, species will continue to become extinct (habitat fragmentation and loss being the main cause of species extinction today). Extinctions will continue without limit until either our species is itself made extinct, or we come to see that our behaviour must change. But even if human pressures on the environment ceased, many species would still be at risk of extinction for some time, as they have been reduced to such small – and therefore vulnerable – populations (Primack, 1993). In comparison, past mass extinctions had finite life spans, the causes of which devastated and then left the biota to recover (if not to reach the same level or range of biodiversity, then to develop an alternative but equally viable level and range of biodiversity).

The current anthropogenic extinction event is clearly unprecedented in respect of this new persistence. Its different nature leads me, once again, to question the validity of extrapolating from past extinction events to the present extinction event. For instance, we could not merely assume that recovery from this new, persistent event (whether with or without humans on the scene) would resemble post-extinction recoveries of the past. Since the events themselves differ, their after-effects cannot be assumed to be the same.

The random model undermines Darwinian fitness

Anthropogenic extinctions fit the random model One difference between the current anthropogenic extinction event and past mass extinctions could be the mode of

---

3 It may be, of course, that humans are not unique in their capacity as aggressive colonisers; it is possible (though not, perhaps, very feasible) that another species might be capable of engulfing all other niches. In that case, we might view the event with similar alarm. However, as far as is known, humans are the only species to have behaved in this way.

species survival. In order to explore this possibility, let me first determine the likely mode of species survival in past mass extinctions.

There are those who argue that all species survival is random, in which case no difference is to be found between modes of survival in any of the extinction events. Acceptance of the random model, though, depends on a rejection of the truly explanatory nature of Darwin's 'survival of the fittest'. If this is seen as a mere tautology – where all who are fit survive and all who survive are fit – then survival will necessarily be random.

Others accept Darwin's 'survival of the fittest' to be truly explanatory, not in a fully predictive sense but in the sense that fitness can be attribute to an individual's particular traits. Gould (1977, p. 42) is of this opinion; and says that:

> certain morphological, physiological, and behavioural traits should be superior a priori as designs for living in new environments. These traits confer fitness by an engineer's criterion of good design, not by the empirical fact of their survival and spread. It got colder before the woolly mammoth evolved its shaggy coat.

So, for those who accept that Darwinian fitness (of which background extinction is a concomitant) is meaningful, species survive because of certain traits of their members that confer fitness.[4]

Yet if fitness is recognised as having an explanatory role in background extinction, it is largely rejected as playing a part in mass extinction events. Gould (1989) argues that mass extinctions are different from background extinctions in that they are random. Mass extinction operates like a lottery, 'with each group holding a ticket unrelated to its anatomical virtues'. Interestingly, though, Gould believes that background extinctions and mass extinctions have more in common than this model suggests. Although he accepts that there is a random element to species survival in mass extinction events, he rejects the truly random model of extinction in favour of 'the different-rules model'. He believes that during mass extinctions 'most survivors get through for specific reasons ... [but] the traits that enhance survival during an extinction do so in ways that are incidental and unrelated to the causes of their evolution in the first place' (p. 306–7). Mass extinctions bring along with them 'different rules' for survival such that:

> the very best of your traits, the source of your previous flourishing, may now be your death knell. A trait with no previous significance ... may now hold the key to your survival. There can be no causal correlation in principle between the reasons for evolving a feature and its role in survival under the new rules (p. 307).

---

[4] The arguments in this book follow those of authors who accept that the Darwinian concept of fitness means something quite distinct from mere survival. See Dawkins (1986); Gould (1977); Ruse (1982); and Sober (1984). There are those who reject such a claim, arguing that one can only determine fitness by survival. Lack of space prevents articulation of the arguments here, but sophisticated variations of this claim can be found in Ollason (1991), Shimony (1989) and Waters (1986).

The key point is that although they are subjected to different rules, the species that survive do so because of specific traits that their members have.[5] Can human-caused extinctions be seen to fit in with this model or are they different?

At first sight, they seem to fit the model. As a result of habitat fragmentation, for instance, the species that survive are those that have specific traits that will fit in with the new rules: having small ranges, good dispersal abilities, rapid reproductive rates and so on. However, unlike in the case of past mass extinctions where the causes – and thus the different rules – were relatively regular in form (usually shifting land masses and/or climate change) extinctions caused by humans take a variety of (often simultaneously acting) forms. Seven are suggested by Diamond (1984, p. 838): '... overkill; habitat destruction by logging, fire, introduced browsing and grazing animals, and draining; introduction of predators; introduction of competitors; introduction of diseases, ... extinctions secondary to other extinctions [and] chemical extinctions'. The resultant 'different rules' have become increasingly complex. It is as if present-day species are being attacked from all sides, with nowhere to run.

A closer look, though, reveals not that the new rules are varied and complex, but that *there are no rules at all.* With past mass extinctions, if a species survived because its members possessed a particular trait – for example small size – we could expect to find other (at least similar) small species also surviving. That is, *reasons for survival were generalisable.* Today that is not always the case. This is because species survival or extinction is ultimately dictated by human preferences, goals and objectives the effects of which are random in a way that past factors never have been. Anthropogenic extinctions may be said to have a truly random pattern of occurrence such that rules simply cannot operate.

Human inconsistency dictates random survival Human preferences cannot be governed by rules because of the variety in such preferences within and between individual humans, and between human communities or cultures. This applies both in the present and across time. But not only is there great variety in human preferences, there is remarkable inconsistency. Consider, for instance, the fact that we hunted the passenger pigeon to extinction because it was small and docile, while we may yet hunt the wolf to extinction because it is large and (reputedly) fierce. It may be argued that this is not odd since the wolf is a threat to us whilst the passenger pigeon was (presumably) good to eat. But this is not satisfactory, because the wolf is also revered.[6] And although pigeons are plentiful in Western cities, they are generally not eaten, even by the poor. We seem to value species' traits in a haphazard, somewhat capricious manner. And the claim about inconsistent or whimsical human preferences is no less valid when we consider plant species.

---

5 It does not matter, here, whether one accepts Darwin's 'survival of the fittest' as truly explanatory; one need only acknowledge the fact that when a species survives (or not), the outcome may be shown to have come about by pointing to the possession (or not) of particular traits. That is, there can be a *retrospective* explanation.

6 Native American Indians reputedly said that when all the wolves had gone, humans would die of loneliness.

Rhododendron, once the popular plant of British Victorian country estates, is now the scourge of conservationists.

Although the direct targeting of species (overkill) is an important cause of extinction, it is generally agreed that the main cause of species extinction today is habitat loss. It is quite clear, though, that the arguments about the randomness of species extinction will apply just as readily to human/habitat relationships. We have preferences for habitats just as we do for species. We are more likely, for instance, to plunder forests than deserts; we have neither the desire nor the ability to destroy (or indeed protect) all habitats equally. Even similar habitats are often treated differently, perhaps for economic, religious or cultural reasons. When decisions to protect or destroy habitats are influenced by such diverse factors as human need, greed and ignorance, great inconsistency cannot be avoided.

The main point of my argument is this: climate change or asteroids landing on the earth were a *consistent* pressure on the species that encountered them; the reason for their survival or extinction (which was to do with the traits their members possessed) *was a transferable or generalisable reason.* Small size was thought to be a generalisable reason for mammal survival during the postulated 'nuclear winter' of the Cretaceous mass extinction. Conversely, human preferences, so often today dictating species survival or extinction, are an inconsistent pressure on species. They cannot be understood within a context of generalisable rules. Since this is the case, the 'different rules' model breaks down. Rules can only come into being if the reasons for survival can be generalised. Human preference is a contingent fact rather than a biological rule.

Human pressures preclude rules for survival At this point, an objection might be raised. It could be suggested that one example of a rule in human-affected environments is: 'species that threaten human food supply are in danger of extinction'. But this will not do. Whilst the statement is in general true, *it is not a rule for survival.* A rule would be required to *differentiate* the chances of species survival/non-survival. All the above statement tells us is about *one* reason for species *non-survival.*

The objection that some species – rats, cockroaches and urban foxes, for instance – seem to have learned the rules for survival in the human-affected environment fares no better. If we consider those various species that are thriving alongside us, we can find no characteristic or trait (such as resourcefulness), the possession of which leads (in most cases) to survival. The nearest we might come to a rule for survival is 'the ability to survive alongside humans'. As we have seen, rules can only come into being if the reasons are generalisable. Yet the fickleness of humans precludes this possibility. Our rule for survival now becomes: the ability to survive alongside *fickle* humans. But to describe human fickleness as a consistent pressure (which is necessary for the formation of a rule) makes no sense at all. By the very *nature* of fickleness *consistency is ruled out.* The idea of a species becoming adapted to fickleness is incomprehensible. And the one adaptive option that might have been open to species – an avoidance of human pressures – is increasingly not an option at all, for the human is now a pervasive influence throughout the globe.

An example will illustrate the point. In Australia, the rabbit (which was introduced there by humans) has adapted perfectly to its new, human-affected environment. It would seem to have learned the new rules for survival. But as the recent situation in Australia reveals, the rules are apparent rather than real. If Australians had their way – and they may yet – the rabbit would be exterminated (Anon, 1995a). It will not do to object that this is merely an example of the rules having changed – just as they would have after an asteroid hit the earth in the past. The point about rules is that they are generalisable. Yet the rabbit alone (or as one of just a few unwanted species) has been targeted for extermination. The reason for the rabbit's demise is too specific. Human preferences are too fickle to allow for rules. Consistency, generalisations and rules have no place in human-affected environments.

Nor does it help to argue that scientific rationality – employed for much of our conservation decision-making – can come to the rescue. As Schrader-Frechette and McCoy (1994) argue, decision-making in conservation is very much based upon human preferences, whether scientific rationality is employed or not. They point out that even apparently objective judgements – such as that of economic efficiency – embrace all sorts of hidden value judgements and assumptions, such as 'the comparative value of biological importance, aesthetic interests, popular demand, anthropogenic needs, and so on'. Objective judgements in a conservation context are usually complex and situational, rendering them little more generalisable than judgements that more explicitly acknowledge preferences.

Similarly, if we consider the survival ability of species that are not deliberately targeted by humans – that is those incidentally affected – little more sense can be made of the idea of consistency. In human-affected environments, species are required to adapt to pressures as diverse as pollution, the motor car, introduced predators and loss of habitat, to name but a few. Clearly, some species do manage to survive. But whether they do or not is nothing short of a lottery. Yet even if it were possible to make some sense of the idea of 'rules for survival' in this limited category, my argument remains unchallenged. It requires only that human preferences – and therefore human fickleness – play an important part in determining the survival or otherwise of species today, not that they determine *all* survival or extinction.

Implications of the random model My argument so far raises the question of whether species that survive despite the whims of humans can be said to be 'fitted' – or adapted – to their environment, for adaptedness, most agree, requires a consistent selection pressure. If (at least some) surviving species are not 'fitted' to their environment, 'fitness' cannot be passed on to offspring and so the concept of evolution is fundamentally changed. To give an example of how this might come about – even in conservation decision-making – consider the following scenario. As Shrader-Frechette and McCoy (1994) argue, when conservation decisions have to be made about whether to save this species over that one, many different factors – often specific to the case – are taken into account. Consequently, it could turn out that due to insufficient finances, popular concern (or a multitude of other possible reasons), the species chosen to be conserved happens to be less well adapted than other species to the non-human biotic, and abiotic environment.[7] Because of a combination of human pressures and lack of help, the species that is better adapted to its non-human environment may subsequently become extinct. Even our well-meaning actions could result in a better-adapted species becoming extinct before its less well-adapted counterpart.

The charge that humans may be undermining Darwinian fitness need not rest on this argument alone. There is another way in which human impact results in maladapted species and this is to do with the *nature* of (many) human pressures on species. Technology drives much change in the human-dominated world. These changes translate into new pressures to which species must attempt to adapt. Yet because of the very nature of certain technological pressures, they rule out the possibility of species adaptation. Take, for example, the motor car. There seems nothing that a species such as the hedgehog could do to take on the might of such an enemy. For a hedgehog, the car may be something that cannot be adapted to; it cannot outrun the car and its usual defence of rolling into a ball is clearly no more successful. And as I discussed earlier, an avoidance strategy is not really an option at all.[8]

Given that, in human-dominated environments the adaptation of species appears to suffer, one might legitimately worry about the possible repercussions on the ecosystem. If the concept of 'health' has any purchase in the natural world (as J. Baird Callicot, 1995, thinks it does) then we might have cause to worry, for a healthy ecosystem must surely be – at least in part – about the adaptedness of the species therein. This point connects with my earlier idea that anthropogenic extinction – as internal to the system – is analogous to disease, which is seen as a failure within the (immune) system. We may have good reason to question the speed, degree, or nature of post-extinction recovery in an ecosystem such as the one that is taking shape under our control. The health issue aside, if adaptiveness is being adversely affected by the current anthropogenic extinction event, then this in itself is a significant difference from past mass extinctions since adaptiveness has been the driving force of evolution since life on Earth first began. Any change after 3.5 billion years must be worthy of note.

---

7 It is not that fitness is only being considered in habitats unaffected by humans; rather, I am suggesting that fitness only truly makes sense where the selection pressures are consistent. As I have shown, there is no consistency in modern-day, human-affected environments.

8 Humans are not necessarily alone in ruling out the possibility of some species' adaptations; other species may be capable of this too. When domestic cats are introduced to islands where the resident species have evolved without ground predators, the latter are sometimes exterminated, as they are unable to adapt quickly enough. Such unevenness between predator and prey could only arise, however, through human intervention that draws on technology. This is not to deny that in the absence of humans some species will cause the demise of other species. It is, however, to assert that there could not be such gross disparity between the survival strategies of two species as to rule out the possibility of adaptation.

Character of impact

Past mass extinctions have all coincided with and been followed by, speciation events. Even though large percentages of species were lost at each mass extinction event in the past, (and after the Permo-Triassic extinction, according to Gould (1989, p. 306) diversity plummeted to 4 percent of its former value) the diversity lost was always regained and indeed went on to increase. So the fact that 'humanity has already raised the rate of species extinction far above the historic rates of species formation' (Ehrlich and Ehrlich, 1981, p. 7) cannot, in itself, suggest that the anthropogenic extinction event is unprecedented, for this was the case during past mass extinctions. It may take a very long time for diversity to be replaced, for it 'took the process of evolution about 50 million years to regain the number of families lost during the Permian mass extinction' (Primack, 1993, pp. 78–9). Or, as the authors of *Biodiversity: the UK Action Plan* (HMSO, 1994) put it, 'there is reason to believe it might take some twenty times longer for the biosphere to recover from the current rate of mass extinctions than humankind has existed on the planet'. Nevertheless, the fact that diversity has increased in a step-wise fashion throughout the history of the Earth despite past mass extinctions must surely be indicative of nature's resilience.

There is, however, a problem with the assumption that diversity will be replaced this time around, and the problem lies in the nature of current human-caused extinctions. Ehrlich and Ehrlich (p. 10) draw our attention to the fact that:

> the same human activities that are causing extinctions today are also beginning to shut down the process by which diversity could be regenerated. Entire new groups of organisms are unlikely to evolve as replacements for those lost if Earth's flora and fauna are decimated now.

The authors do not expand on this point but the theme is taken up by Vermeij (1986) who gives a speculative explanation. His argument rests upon the claim that current habitat fragmentation, although not a new phenomenon, is fundamentally different from past fluctuations in habitat size, which, he acknowledges, were also responsible for mass extinctions (p. 30). He notes that according to the classical – and popular – view, speciation is promoted by 'geographical isolation coupled with a distinctive regime of selection' (p. 42), drawing our attention to the paradox that both speciation and extinction are favoured by the same conditions.

Although it may seem that human habitat fragmentation should favour speciation, Vermeij argues that it may not do. The reason, he suggests, is that unlike in the aftermath of past mass extinctions, current populations of species are unable to expand because of continued pressure on their habitats. As we have already seen, current human pressures on habitats are persistent; arguably uniquely so. Vermeij argues (p. 42) that:

> Because more genetic variation is preserved in an expanding population than in a declining one, the successful incorporation of favourable genetic alterations is more likely in an expanding population than in one which is stable or contracting. Population expansion may be impossible in most instances unless the population's habitat is expanding.

This statement immediately rings alarm bells, for in most cases protected natural areas are decreasing rather than increasing in size; in some cases whole areas are de-designated, while in others, protected areas are being bisected by roads, or encroached upon from the edges, thus fragmenting them or whittling them down little by little. Indeed, it seems that for certain organisms there is a minimum area that is necessary for the process of speciation to take place. For example, 'for small mammals, the smallest islands (Cuba & Luzon) on which a single species is known to have given rise to two species are 100,000 km^2' (Primack, p. 30). But not only are habitats now small and either remaining so or decreasing in size, resulting in (at the very least) reduced speciation, speciation is still further reduced by the fact that we have lost vast stretches of habitat to human settlement. This use – which is increasing apace – is largely incompatible with the evolution of new species. These two factors reduce *total* diversity – the variety of species in a geographic area – (as opposed to within-habitat or between-habitat diversity) the importance of which is emphasised by Norton (1986).

However, the sort of speciation we have been talking about – allopatric speciation – in which speciation occurs as a result of a geographical barrier is not the only kind of speciation. Other kinds include polyploidy (a common form of – particularly plant – speciation) and sympatric speciation – that which occurs in the same place as the parent species among, for example, insects. Neither of these types of speciation requires expanding habitats. They can also occur very rapidly. The question arises, then, why worry about decreasing habitats or indeed extinction?

While it may be true that species could, theoretically, continue to be replaced through these kinds of speciation despite reduction of natural habitat, this would be unlikely to placate conservationists. To begin with, although speciation of this kind does not require expanding habitat, it does require *some* habitat. If we continue in the direction we are heading, we may not one day be able to accommodate even this meagre requirement. At the very least we are in the process of reducing habitat in which such speciation can take place. Further, although new species may be formed, they will not – for many of us – be adequate replacements for those lost. They will consist mainly of plants, insects and other small species; there will be no replacements for the elephants of this world through these systems of speciation. Further elucidation is best left to E. O. Wilson (1992) who writes (pp. 67–68) that:

> new species are usually cheap species. They may be very different in outward traits, but they are still genetically similar to the ancestral forms and to the sister species that surround them ... Pairs of newly created sister species are often so close in their diet, nest preference, susceptibility to particular diseases, and other biological traits that they cannot coexist ... local communities are not enriched by the presence of both.

Theoretically, new 'cheap' species could, of course, evolve over time, moving genetically further apart from one another. However, such evolution would be severely limited by the reduction of suitable habitats as explained above. At best, we may still see some genetically diverse species evolve but these will be restricted to plants and very small animal species such as insects. The evolution of something like a panda not only requires luck, trial and error and a vast amount of time, it also requires suitable habitat. If we are concerned about the sort of speciation that will take place in the future, then the reduction of habitats is relevant.

The fact that there is a relationship between the size of a habitat and speciation (i.e. there being a minimum size that allows speciation to occur) is sufficient to explain the tension between the idea that speciation is unlikely to occur in artificial reserves and the fact that speciation is particularly rapid on islands, given that, in the words of Primack (1993, p. 88) 'reserves can be viewed as *habitat islands* in an inhospitable 'sea' of unsuitable habitat'. There is, however, a further point that helps to show how nature reserves cannot hope to produce new species in the way that islands do and it centres on our habit of species introduction. Vermeij (1986, p. 43) argues that on true islands, the biota was insular and had relaxed biological selective regimes whereby species evolved interesting and unique adaptations. Introduced species have, however, 'rendered the selective regime ... more like that of continents'. Vermeij sees us as 'homogenizing the selectional environment of organisms and inhibiting the evolution of novelty...' (p. 44). If species introductions are having this effect on existing islands, it seems unlikely – given the greater ease with which this can take place across land, and the probable continuation of this practice – that refuges can ever really become established as true islands. And further, if species introductions have a homogenising effect, then colonisation – which must occur to replace those species that will naturally become extinct – will be of species more similar to those that already exist on the refuges than would have been the case on true islands. And this is so, not only because of introductions, but also because habitat fragmentation and species extinction continue unabated, reducing the pool of diversity all the while. If colonising species resemble (or are the same as) those species already there or recently lost, then they do not represent 'new stock' from which divergence can take place. Colonisation will, of course, be particularly important in island-like reserves for it has been shown that the smaller the island the greater the extinction.

It may be, then, that anthropogenic extinctions are quite different from past mass extinctions in that subsequent speciation will be of a very limited kind, if it occurs at all. A further difference follows from this and that is the reversal of a general (step-wise) trend towards increasing diversity. It looks as if it is indeed the case, as was suggested earlier, that we cannot extrapolate from past to present mass extinctions in terms of recovery.

Conclusion

I have put forward four arguments to support the claim that there are important differences between the current anthropogenic mass extinction and past mass extinctions. They lead to the conclusion that the current extinction event is, indeed unprecedented. Recovery from this extinction event cannot be assumed on the basis of recovery from past extinction events.

The four arguments have been offered individually – and can certainly be seen as distinct from each other – yet they are not entirely separate in their significance. Together, they have a cumulative effect that serves to formulate a robust position. Each argument plays its part, both individually and collectively, in bringing us closer to the conclusion that the current extinction event being brought about by humans is significantly different from other mass extinctions.[9] And it is the idea that recovery is in question – in terms of diversity alone or even any recognisable form of recovery at all – that is the ultimate focus of concern. This may be an anthropocentric concern (we wish to continue as a species and fear this may not be possible in a radically different world), but it may equally be an ecocentric concern (we recognise that other species, mountains, rivers and so on are entitled to exist in their own right).

There would seem to be sufficient evidence to conclude that anthropogenic mass extinctions are sufficiently different in kind from other mass extinctions to warrant different judgement about their significance. This different judgement does not necessitate a normative conclusion. That is, we cannot conclude that *therefore* extinction is undesirable, for first one must value a diverse and thriving ecosystem. However, arguments for the value of biological diversity have been provided elsewhere (e.g. by Norton, 1987), while what has been provided here is evidence of the *coherence* of the conservation assumption that 'the untimely extinction of populations and species is bad'. That is, it makes sense to suppose that two extinction events can be differently evaluated only if there is some significant factual difference between them. And so by revealing the true nature of the current extinction event, I lay the groundwork for the belief that such extinction is indeed undesirable. The gut-feeling expressed by conservationists – that extinction is wrong because it destroys beautiful and interesting kinds of lives – is open to the criticism that sometimes such destruction will be viewed as a *good* thing (e.g. in the case of the smallpox or AIDS virus). This leads to the necessity for a highly problematic explanation of where to draw the line between those kinds of lives that are, or are not, of value. The arguments, here, would not be so affected, because they consider current anthropogenic extinctions not as individual occurrences, but rather, as a single *event*. That is, my argument is not designed to show why every single extinction is necessarily bad, but why the current anthropogenic extinction event, as a whole, is so undesirable and unwelcome.

If it is so that there is a sound basis from which to argue that species extinction is undesirable, are we to suppose that there is no need to question the current conservation emphasis upon the safeguarding of species? I will show later (in Chapter Six) that there are good reasons to suggest that concentrating too closely on species is a mistake. But with respect to my particular arguments in this chapter, there are reasons of another kind to suppose that conservation ought not to focus so heavily on safeguarding species.

We have seen that conservationists are fickle in their decision-making and action such that the impact they have on the natural world could be affecting the adaptiveness of species. Adaptiveness – as noted earlier – has been the driving force

---

9 One further difference between anthropogenic and other extinctions is, of course, the fact that we can choose not to cause extinctions. Some might argue that this is the most important difference between them. However, the fact that we are able to choose tells us nothing about the difference between the extinction events themselves or about the desirability (or otherwise) of making a particular choice.

of evolution since life on Earth first began; as such it is an integral part of the evolutionary process. Without adaptiveness, then, two primary underlying values of conservation – the evolutionary process and its product of biodiversity – cannot be sustained. Adaptiveness, then, must also be an underlying conservation value. Given that the conservationists' inconsistent treatment of populations and species poses a threat to the value of 'adaptiveness' it would seem reasonable to move the focus of concern away from 'safeguarding species' – with its emphasis on particular entities – and towards the wider concern of 'safeguarding the opportunity for speciation' a concern that would demand, in particular, the safeguarding – and expansion – of habitats. Such a focus would be more in tune with conservation's values, not only because it would remove emphasis from an area where human fickleness adversely affects one of conservation's own, very basic values – adaptiveness – but also because, as we saw earlier, this alternative focus is necessary if we are to bring the current anthropogenic extinction event in line with past mass extinction events in terms of recovery.

Chapter 2

Rarity

Introduction

In Chapter One I considered whether the conservationist is justified in assuming that the current anthropogenic extinction event should be differently evaluated from other kinds of extinction, an assumption that lies behind the belief that current human-caused extinctions ought to be prevented. I found that there are good reasons for supposing them to be very different in kind, reasons that, if biological diversity is valued, give great cause for concern and show that such extinctions are highly undesirable. Conservationists should seek to prevent them, but the prevention measures ought to focus primarily on the provision of opportunities to actively reverse the extinction trend; that is, to allow for speciation.

Nevertheless, if we accept that biological diversity is a conservation value, it seems that we must also accept that the extinction of biological types (species, subspecies, populations) *where replacement may not occur*, is to be avoided. Conservationists, then, seem justified in concentrating policy measures, as they do, upon the prevention of such extinctions. Reading through virtually any conservation policy document, you will notice that an important aspect of such anti-extinction measures is an emphasis upon rare species. It may seem obvious that if biological diversity is valued, when its loss is imminent – that is, where a species is *threatened* and particularly where there is little hope of replacement through speciation – there is every good reason for conservation policy to concentrate upon preserving that rare or threatened species. The importance to conservation of rare species is accepted without question: rare species are the proper concern of the conservationist.

A corollary of conservation's emphasis upon rare species and other assemblages, are the assumptions, firstly, that the perpetuation of a rare species ought to be prioritised over the survival of some members of a common species, and secondly, that rare species are more valuable than common species, subspecies and the like. By way of an example, at Arnside Knott in Cumbria, the National Trust is restoring a rare habitat – limestone grassland. This in turn supports a rare butterfly, the high brown fritillary. The result is the destruction of the more common wooded habitat, home to an array of common species. My aim in this chapter is to discover whether it is so that, in the conservation domain, rare species always deserve such preferential treatment at the expense of common species.

What kind of value does rarity have?

In order to discover whether rare species should take precedence over either common or abundant species in conservation action, I first need to determine what kind of value rarity has. For example, does rarity merely amplify the existing value of a thing or is rarity itself a source of value? The belief most widely held (such as by John O'Neill (1993, p. 183)) is that rarity is an amplifier of value. O'Neill writes that 'since everything is rare under some description, rarity as such cannot confer value on an object ... it is best understood as an amplifier of value'. Whilst initially convincing, a closer look at this claim reveals it to be problematic.

Rarity as an amplifier of value?

The first difficulty with O'Neill's claim is with the idea that 'everything is rare under some description', for it commits us to the view that consequently, everything is rare. But to imply that everything is rare is simply a distortion. Whilst it may be true in some kind of technical sense that everything is rare under some description, such a claim has little real meaning or content. Just as, in a technical sense, everything is large under some description, we could not be persuaded that consequently, everything is large, so too with rarity. We are perfectly well able to differentiate between that which is rare and that which is not; we have a good working understanding of the concept.

The problem with O'Neill's claim that 'everything is rare under some description' is the assumption that all descriptions are equally valid, that any possible reason for ascribing rarity is as good as any other. This, however, is not the case; just any description will not do, it must be a description *of the right kind*. There are, in fact, conventionally recognised constraints in our talk of rarity.

To begin with, rarity is different from uniqueness. It may be that something that is unique may also be rare: the last surviving panda, for instance. However, it is not so that something is rare *in virtue of* its uniqueness, for uniqueness is extremely commonplace; there are billions of unique things. Furthermore, unlike rarity, 'uniqueness does not admit of degrees (no one can be more unique than anyone else)' (Gunn, 1980; p. 32). Nor can it be, as O'Neill seems to suppose, that rarity requires only 'the sharing of some description with very few other items'. A jar of Marmite is not made rare by virtue of the fact that it is one of only three I have ever bought. In order for a description to count as a description of rarity the shared features must be of *significance* to us. Descriptions that are of no significance to us, we might call 'contrived' or 'accidental'. Primarily, the thing that distinguishes a significant description from a contrived description is that the former is set within a conceptual framework in which 'norms' of distribution are already established. Something can only be rare against such a backdrop. Since we have no conceptual norm regarding the average number of jars of marmite bought, it is clear that this description is of no significance to us. For rarity to have meaning, there must be something significant about the thing being rare in that place. This will only follow when both the thing that is being described, as well as the relationship to the place in which it is found are significant to us. To say that stamps are rare in Opera Houses is not taken seriously because their being in these places (or not) is of little interest or significance to us. If we were to say, instead, that stamps are rare in France, this would be seen to be of great significance, for their rarity would make life very difficult for the French. We have a conceptual framework regarding the normal distribution of stamps around the world, and so talk of their rarity is meaningful.

The norms that we establish within our conceptual framework – and which must be in place before a thing can meaningfully be rare – apply to categories of things – that is, categories of things that we consider significant: these might be, for example, Van Gogh paintings, abilities, diseases or species. Rarity, then, is a distributional claim, requiring reference to others in the same category. Rare things are repeatable; they have a degree of constancy and so are recognisable as belonging to a specific category. (The only kind of example in which something unique is also rare is one in which the unique thing belongs to a specific meaningful category: the last panda belongs to the category of 'panda species', while the unique Van Gogh belongs to the category 'Van Gogh Paintings'.)

I could summarise by saying that the only kind of description that will confer rarity upon something (that is, is a description 'of the right kind') is one in which the thing described belongs to a category that is significant to us, either in itself and/or in terms of its relationship to the place in which it is found. When a description is significant to us a conceptual framework is in place as regards the normal distribution of the thing described. I suggest, then, that O'Neill is wrong to suppose that 'everything is rare under some description'. To suppose so, is to misuse the term 'rarity'. Some descriptions will simply not do. Since O'Neill's conclusion that 'rarity as such cannot confer value on an object' relies upon this premise, it is considerably weakened. Perhaps rarity does confer value on an object.

Perhaps, though, O'Neill is implying that we are mistaken in thinking we can recognise rarity when we come across it, and that he is exposing the truth of the matter in claiming that 'everything is rare under some description'. That is, he may be criticising those conventionally recognised categories. However, if, as O'Neill supposes, everything is rare (in some way), if rarity amplifies value – as he also believes – then the value of everything is amplified in relation to its original value. This means that we are unable to differentiate, either before – or after – amplification of value, between rare and non-rare things. To put it another way, if everything is rare under some description, then rarity is an empty concept.

The third difficulty with O'Neill's argument is one of logic. To say that 'everything is rare under some description' does not commit us to the conclusion that rarity *cannot* therefore, be a source of value. To see this, we need only replace 'rare' with 'good'. Although it may be possible (technically) to say that everything is good under some description, it does not follow that 'good' cannot be a source of value. Clearly it is. This is not to deny that different descriptions of 'good' will be differently valued, it is merely to point out that even if O'Neill were right in supposing that everything is rare under some description, this would not rule out the possibility that rarity could, all the same, be a source of value.

There are, then, various problems with O'Neill's argument that rarity cannot be a source of value but must instead be an amplifier of existing value. Clearly, though, we cannot reach any definite conclusions regarding the value of rarity. To begin with, I have considered only one line of argument for the claim that rarity merely amplifies value. But further, I have only shown that because of the problems with this particular argument, the way is left open for the possibility that rarity may be a source of value. It could be, for instance, that other arguments claiming that rarity amplifies existing value are more convincing. Gunn (1980), for example, puts forward an argument for the claim that we do not value paintings for their rarity alone: 'Unless we have some reason other than mere rarity, we do not value paintings, whether they be many or few' (p. 30). He makes a similar claim for the amplifying value of rarity in the case of stamps: 'If there were not a demand for stamps by collectors who find them interesting *per se*, rare stamps would be as worthless as common ones' (p. 31). I might point out that the concept of rarity cannot be separated from, is in part understood as, a category of things that has significance to us, as established above. Yet even without this response, Gunn's argument may be unconvincing with respect to the conservation context. He makes the observation that if, in the case of art works and collectors items, 'rarity is a factor only in increasing and not in creating value [then] it is hard to see how the preservationists can make a case for preserving rare animals merely because of their rarity' (p. 30). However, Gunn may not be justified in making the jump from paintings and the like to species. It may be that we cannot make any assumptions about rarity and species on the basis of our attitude towards other rare things. We need to consider rarity specifically as it relates to the natural world.

Rarity and species

The idea that rarity merely amplifies value does not seem to tally with the way many conservationists appear to think about rare species. It does not always seem to be the case that conservationists already value the thing that, when rare, they consider to have value. The smallpox virus, for example, is seen to have negative value, and we demonstrate this by destroying it whenever we can, conservationists being no exception. Yet Ridley (1994, p. 15) claims that most environmentalists would consider the extinguishing of the virus to be a mistake. So even though it has negative value in normal circumstances, the impending loss of this organism – or its rarity – would seem to confer value upon it.

For a more tangible, though hypothetical example, consider the magpie, one of Britain's most common bird species, and increasing in many areas. This bird, though beautiful and intelligent, is much persecuted by gamekeepers and increasingly losing favour with the public who accuse it of eating garden bird eggs and chicks in ever-greater numbers. Many conservationists consider this bird to have negative conservation value. Imagine, now, the outbreak of a disease specific to the magpie such that the majority are wiped out. Conservationists would no doubt be falling over one another to protect what remained of the now rare species. (This is certain to be so if the magpie became globally rare, but it is also probably true if the disease struck British magpies only; indeed, national boundaries are one important parameter for designating a species as rare.) Since it would only be their density that had changed, it would look as if it was rarity that conferred value upon the magpie. Is it really so that rarity is valued in itself or is this an illusion?

It could be an illusion. There may be a way of reconciling the above attitude with one in which species are valued, and where this value is amplified by rarity. 'Species' and 'individual' may be being conflated: conservationists might value the magpie as a species but attach negative value to the individuals that are thought to endanger garden birds. This may, however, be harder to explain in the case of, say, the smallpox virus; it may not be easy to see its species value. All the same, this value might be understood to be in terms of biodiversity and future (as yet unknown) uses. If species existence is valued, then when that existence is threatened by rarity (that is, when the species is endangered) rarity might be viewed as the amplification of existing value. The idea that a rare thing can have value only if the thing in question already has value is plausible if we think of rare things such as cyclones and mass murderers: though rare, we clearly do not value these things. But it could be that part of the reason we do not value them is because there is no sense of them ever being *endangered* (their occurring does not depend on previous occurrences). Future living organisms, on the other hand, depend for their existence on the survival of present living organisms. Once a species is lost, it is (with our present level of knowledge) lost forever, and it is this sense of impending loss that appears to give rarity – in relation to living organisms – its bite. Could it be, then, that rarity does, after all, merely amplify the existing value of species and that the search for rarity as a source of value is over? Though at first this may seem to be the case, a more thorough analysis of species rarity suggests a quite different conclusion.

Species rarity is to be found in two different forms. The first of these, we might call 'artificial rarity'. This is the kind already spoken of, where the species is to some degree, endangered. The giant panda and the white rhinoceros fall into this category. Since conservationists' efforts are directed towards *ending* such rarity, it is clear that, in these situations, rarity cannot be a source of value. I suggested, above, that in cases where species are endangered, rarity might be considered to be an amplifier of existing value. There is, however, no reason to suppose that rarity amplifies value either. Rather, it is much more plausible to suppose that rarity merely acts as a *pointer* to the existing value of the species by highlighting its endangered status. We value the species and so when it is in danger of extinction, more effort is required to maintain its existence, or rather its viability, than is required in the case of common species. Just as, when a person is ill and they are given extra care and attention this does not mean they are any more *valuable* than their healthy counterparts, so too with species: rare species require more attention than do common species but this does not mean they are more valuable. Of course, a rare species *may* be more valuable than a common one, but not for the reason of its rarity.

Since conservationists are largely concerned with 'artificial' rarity (the fight against species extinction), for the most part the rare species they deal with, though prioritised, are not more valuable than any other species. To suppose that they were, would be to give them preferential treatment over other species, at all times. To appreciate that this is not the case, it is worth considering the following thought experiment in which conservationists have to make a choice between one of two policy decisions, A or B. Policy A leads to the extinction of species A which is very rare and highly endangered, whereas policy B leads to the extinction of species B which is abundant and flourishing. Assuming that the only significant difference between them is their relative abundance, it would seem that there are strong arguments for policy A, given that what is lost as a result is something very likely to be lost in any case. This demonstrates that rare, endangered species are not, by virtue of their rarity, more valuable than common species; conservationists' prioritisation of species is largely pragmatic. (Indeed, this policy decision would no more indicate that common species were more valuable than rare species.)

Although 'artificial' rarity is the form of rarity most usually dealt with by the conservationist, there is also what might be called 'natural' rarity. Such species are not endangered in the sense that their numbers are approaching or have fallen below a level of abundance that is thought necessary for their viability. For example, *Amorphophallus titanum*, a member of the arum family of plants found in Sumatra, is extremely scarce, with individual plants widely scattered; furthermore, it flowers for only three days out of every thousand, during which time it has to be pollinated. It succeeds in attracting its pollinators – tiny sweat bees – from great distances by giving off a stench of dead fish from its seven-foot tall bloom (Flowers, 1995; p. 50). It is, in other words, supremely adapted to its rarity. Rarity, then, may be considered part of this species' value.

It may be objected, though, that 'naturally' rare species are still at risk from extinction, which creates a tension between the idea of rarity as part of a species' value, and the need to reduce rarity in the interests of preventing extinction. In one sense, of course, any rare species is more endangered than any common species, for they are more at risk from, say, catastrophes such as floods or volcanic eruptions, especially if the species is restricted to the location in which these things occur. However, by far the greatest risk to rare species is human interference, and the moment humans have a significant adverse effect upon a rare species, it is no longer 'naturally', but 'artificially' rare and so in need of measures to increase its numbers. In any case, it does seem to be true that 'naturally' rare species are more resilient than 'artificially' rare species as Gaston (1994, p. 162) observes:

> A distinction can be drawn between species which would be rare in the absence of human activity ... and those which are rare because of it. It seems probable that the former may be less prone to extinction as a consequence of their rarity than the latter, being more likely either to have evolved adaptations which reduce their vulnerability or being survivors of selective extinctions because they possess such characteristics.

Furthermore, threats facing 'naturally' rare species – such as floods and storms – are not of a *persistent* kind as the threats facing 'artificially' rare species usually are (as argued in Chapter One). Since rarity is part of what a species such as *Amorphophallus titanum* is, since it is part of its make-up, then if the species is valued, its rarity must, necessarily, be valued too. Here, rarity does not act as a pointer to existing value as it does in the case of 'artificially' rare species. Nor does rarity amplify value, for rarity is not something that can be added on to or withheld from such a species; it is part of it, one of its characteristics. However, neither is it true that rarity, here, is a source of value, for again, in order for this to be so, it must be possible to separate or differentiate rarity from the species itself; yet this cannot be done, for part of what such a species is, is rare. All that can be said is that rarity is part of the value of a 'naturally' rare species.

To say that rarity is part of the value of 'naturally' rare species is not to say that such species are more valuable than non-rare species; these other species, if valued, will have characteristics of their own that contribute to their value. For example, in those cases where we value abundant species, part of what we value about them is their abundance. The American bison is valued as an animal that existed in huge herds and we regret the disappearance of this spectacle. Similarly with the passenger pigeon: part of the reason it is so often quoted as an example of an extinct species is because of the enormity of its number prior to human exploitation. And closer to home, we value the starling as a bird that gathers in huge winter flocks that swirl across the sky in ever-changing patterns. Likewise, with common (as in widespread) species: the British robin is valued, in part, because it is such a familiar and reassuring sight. So where rarity is a species' natural characteristic, there is no reason to suppose that it is more valuable than any other species. Rarity is merely part of what we value about a 'naturally' rare species.

Rarity as special?

So far, I have considered rarity with respect to 'artificially' rare species (wherein rarity was found to be a pointer to a species' existing value) and 'naturally' rare species (wherein rarity was found to be a part of a species' existing value). In neither case, then, is rarity a value in itself. Yet it does seem, when considering the concept of rarity, that there is something *special* about it. It seems plausible to claim that there is something special about a thing existing in a limited way.

This specialness, or value, might be said to have its roots in the following:

- Rarity contributes to *diversity of dimension*. Rarity is part of a continuum from non-existence to commonness or abundance. Where rarity offers an alternative to non-existence, it clearly contributes to diversity. As a blanket alternative to commonness, rarity would not offer further diversity, but rarity in *addition* to commonness clearly does. The existence of rarity, then, helps to guard against a kind of homogeneity, a tendency towards similarity in our environment with a concomitant similarity of experience. If we value diversity of experience then we must value rarity.
- The kind of experience that rarity gives us is valuable in its 'limitedness'. We find it thrilling to encounter, for example, a rare species that has seldom been seen by others, perhaps because of its secretive nature or inaccessible habitat. Equally thrilling might be the sight of a species rarely found 'in these parts' but which has, perhaps accidentally, ventured there. Migratory birds that veer off course are one such example. The rarity of seeing a rare thing is part of the value of the experience.
- The rarity of a thing seems to enrich our experiencing of it. When we encounter something rare, we seem to become more aware of its inherent qualities: we notice its colour, shape, texture, movements and so on, which in similar but common things, we often remain unaware of. In just this way, we may become fascinated with the contours and colours of rare gemstones.

My idea is that once we encounter a rare thing we become more conscious of its inherent qualities because we already value its rarity; we often have a more intense experience. And this idea seems to hold true with species. Considering the opposite angle, starlings, pheasants and magpies are very attractive birds, but their beauty often seems lost on us, arguably because they are so commonly seen. Were they less common, we may be more inclined to notice their beauty. As a thought-experiment, it is worth imagining how the experience of watching a starling would be likely to change, were you to discover that this bird is, after all, a rare subspecies. Indeed, it is not only that rarity draws our attention to the rarity's inherent qualities; rare things often take on *extra* qualities. They may be thought of as magical, allegorical or symbolic; they may engender interest, passion, awe or respect. This has been seen to be true of rare (or rarely experienced) species over the ages. Consider, for instance, the albatross. Rarely seen, it has been thought, by sailors, to have spiritual significance. When rarity is considered, as *an experience*, it does seem to be special. Might it be, then, that rarity is, after all, a source or amplifier of value?

The above thoughts do seem to suggest that there is something about rarity in itself that is valuable, a value that is tied to but precedes the thing to which it refers. The kind of value that rarity has may be likened to the values of 'love' or 'beauty'. These we can describe as amplifiers of value in that they can enrich our experience of life, but there is also a sense in which they are sources of value. Human nature is such that love and beauty – and arguably rarity too – are things that we strive to experience. They are values towards which we are drawn; they are worth having in themselves. However, rarity belongs to a category of values that may not always lead to a valued outcome. That is, the product, as it were, of these values depends upon the valuer. Rarity, for example, will sometimes engender interest of a less than admirable kind. It may pander to our egos rather than enrich us. In that it is 'elitist' by nature, it can bring out the very worst in us, such as the fanatical egg-collector who will go to any lengths to gain another bird's egg for his or her collection, regardless of the harm done to the un-hatched chick, the parent birds, the species or the ecosystem. Similarly, we greatly value love, yet in certain individuals it can engender undesirable behaviour in the form of jealousy, paranoia, possessiveness and so on. And again, food that is a source of value not only as vital sustenance but also as a way of enriching our lives may, for some, become an obsession, leading to, for example, gluttony, anorexia nervosa or bulimia. Innumerable other examples of things we value are abused by some valuers, such as patriotism, discipline, sex and money. Such values differ from, for example, the values of 'health' or 'happiness' that are more clearly a good in themselves. We might best describe values such as 'love', 'beauty' or 'rarity' as *potential* sources of value. That is, they have the sort of qualities that tend to enrich our lives and promote further value. Rarity *tends* to promote a richer experiencing of the rare thing, appreciation of beauty, wonder and respect.

Since it is clear, now, that rarity is only potentially a source of value, we can see that there is no difficulty in claiming that it is a source of value in some cases ('natural' rarity) but not others ('artificial' rarity). One possible objection, however, might be that the specialness of rarity applies not only to 'naturally' rare species, but also to 'artificially' rare species. This would contradict the earlier claim that 'artificially' rare species are *not* a source of value. It may, indeed, appear that endangered species engender those positive experiences described above, suggesting that such species may be a source of value. However, there are good reasons to suppose that this is not, in fact, the case and for the following reasons.

First of all, part of the value of rarity, as we discovered, is that it contributes to diversity of dimension – that is, that rare species contrast with common or abundant species. 'Artificially' rare species, in being endangered, are likely to be ephemeral and unsatisfactory contributors. Either they will disappear altogether or, less probably, their population will recover such that the species is no longer rare. Yet further, as a contribution to diversity of dimension, rarity need only *exist*, there is no merit in there being many examples of rarity. So the examples of 'natural' rarity that exist are quite sufficient to satisfy our desire for a 'rarity dimension'; examples of 'artificial' rarity are superfluous. Yet it is a stronger point than this. The more examples of rarity there are, the less special it is and therefore the less enriching the experience, so the burgeoning of rare species (which is largely of 'artificially' rare species) must be seen in a negative way. Rarity ought to remain rare! 'Artificial' rarity, then, contributes nothing towards diversity of dimension, and indeed works against it. This might provoke the response that if this is so, then all that is needed is one single example of a 'naturally' rare species to fulfil the need for a rarity dimension. It seems reasonable to suppose that a *dimension* requires more than a single example, but more pertinently, there is already a presumption against promoting 'artificial' rarity or extinction, which means that 'natural' rarity ought to consist of the range of species that currently are 'naturally' rare.

The second point concerns the worry that 'artificially' rare species will engender the feelings of pleasure described above in relation to 'naturally' rare species. It is true that we may, indeed, be likely to become more aware of 'artificially' rare species' inherent qualities. But rather than such observations giving rise to pleasurable experiences, they are more likely to produce the opposite response. In place of wonder, respect and awe – those feelings engendered by 'naturally' rare species that are *winning* their battle for survival (as we saw, they are supremely adapted) – 'artificially' rare species, that are *losing* their battle for survival will tend to engender negative feelings such as anger, frustration, sadness, even shame. And it is not just the case that the negative outweighs the positive; it is, rather, that the *idea* of 'artificial' rarity is approached in a quite different manner. Whether or not a rare species engenders pleasure is tied in with the valuer's knowledge of its history.

And thirdly, even if it were true that 'artificially' rare species can engender some kind of wonder, this does not create a problem for the main thrust of my argument: that rarity, *sometimes* is a source of value. All the same, the claim that 'artificially' rare species can promote some positive experiences would not be proof that 'artificial' rarity is a source of value, for it is quite possible to value a rare thing whilst not necessarily supporting its rare state having come about. Just as we can value the way that a disability seems to bring out the best in a person, without necessarily wishing for everyone to be disabled, so too might we appreciate certain qualities in a species that have been highlighted by its rarity, without finding value in the state of 'artificial' rarity itself.

Having dealt with one problem relating to the claim that rarity is sometimes a source of value, I must now tackle another, and it is this: how can the claim that rarity is special and a potential source of value be reconciled with the earlier claim that rare species are no more valuable than other species? Earlier, when discussing the way in which rarity is part of the value of a 'naturally' rare species, I suggested that other species have qualities that are part of their value, and that there is no reason to suppose that any species' qualities are more valuable than any others'. A similar argument applies here: though rarity is special, other modes of being are special too. As I asserted earlier with the bison and starling examples, abundance is often a source of value. Furthermore, non-rare species will have other things about them than mode of being that are special, such as, say, an interesting (or indeed simple) life history, or an intricate relationship with another species. It may even be that a non-rare species engenders the same response as does a rare species but for a different reason. Sometimes, for instance, the intricate way we know a common species may result in an experience that is as enriching as one involving a rare species. Clearly, there is a range of values that relate to species and there is no reason to suppose that we ought to be able to place these values in a *hierarchy*. We value health as well as happiness, but would find it difficult to say which we found more valuable; certainly, we would be unlikely to reach a consensus. Similarly, we value both complexity and simplicity, yet it would be impossible to place their value in any order; each is special in a quite different way.

But it is not merely that the different values of different species cannot be placed in a hierarchy; many values are interdependent. Rarity, for one, depends upon its opposite number, namely 'abundance' (or 'commonness'). 'Complexity' and 'simplicity' are similarly interrelated. One would be neither meaningful nor valuable without the other. And since each is at one extreme of a continuum, their value equally depends upon the states between the two. Consider, for example, that 'rarity' needs 'commonness' to temper its excesses: a species that was too rare would be non-viable. It seems, then, that it is perfectly acceptable to claim that rarity is a source of value and special, without having to conclude that rare species are necessarily more valuable than other species.

Let me summarise my argument so far:

- The rarity of 'artificially' rare species is not something we would wish to preserve so it cannot be a value at all. Such rarity merely points to the existing value of the species.
- The rarity of 'naturally' rare species, as part of their makeup, is something we would wish to preserve, provided we valued the species.
- There is something special about a species existing in a limited way and so rarity is a (potential) source of value.
- However, rarity is just one of many values that relate to species. Rare species, then, cannot be considered to be more valuable than other (for example, common) species.

Keeping in mind this clarification of the meaning and status of rarity as it relates to species let us consider how these findings map onto conservation policy.

Rarity and conservation

In the vast majority of cases in which conservationists concern themselves with rare species, the species is 'artificially' rare; it is, to some degree or other, endangered. In most instances, then, conservationists are working to counteract rarity. The rare species they have dealings with are not, as we have seen, any more valuable than non-rare species. It is simply the case that they are in need of help if they are to remain extant. It is important that this seemingly self-evident point is made explicit, because there would seem to be an implicit assumption that because conservationists focus their efforts particularly on rare species, rare species are of greater conservation value than non-rare species. Once it is realised that this is not, in fact, so, there is more room in conservation thought and action, for valuing common species. This realisation may make little practical difference when the conservationist is faced with fast-disappearing species (that is, the conservationist will still, no doubt, have to concentrate on saving these species). However, it will make a difference in the attitude of the conservationist – as well as of those who look to the conservation rationale for guidance and inspiration – towards common or abundant species. The prevailing assumption is that since conservationists focus upon rare/endangered species then common species are of no consequence to conservation. This will have to be rejected as a misconception.

Whilst conservationists do, for the most part, encounter rarity in relation to 'artificially' rare species, some of their work does involve 'naturally' rare species. Recalling the example of Arnside Knott, it would seem that the National Trust is seeking to increase the number of high brown fritillary butterflies in this area, yet this butterfly *has always been rare here*; that is, it is 'naturally' rare. Thus, the National Trust is seeking to increase the butterfly population, not until it reaches a former density, but rather, *beyond* any former density. I have argued that it is perfectly acceptable for conservationists to intervene to alter the status of 'artificially' rare species. Is it equally acceptable for them to alter the status of 'naturally' rare species, given that such rarity is, as I have shown, not only part of the value of such a species but a special occurrence, a potential source of value? In other words, is it acceptable for the conservationist to make a rare species common? And, if this is not acceptable, then ought the conservationist to *prevent* a rare species becoming common? How important, then, is rarity when it occurs naturally?

In order to answer this question, I need to ask another: what do we mean, exactly, by 'naturally rare'? For example, if a 'naturally' rare species increased, in what sense could it be thought of as having been 'naturally' rare in the first place? And might it, in its new state, have to be thought of as 'artificially common'? And if this were so, it may reasonably be supposed that a species would lose conservation value if it were moved from the category of 'naturally rare' to that of 'artificially common' (assuming that conservation values the natural above the artificial).

Intuitively, it seems fair to consider a species 'naturally' rare if it occurs in a rare mode prior to deliberate interference. Such a species may or may not be termed 'artificially common' if interference causes it to increase. Regardless, there are good reasons to claim that 'artificial commonness' need not be perceived in the same negative light as 'artificial rarity' – that is, in terms of having no conservation value. Certainly, in the sense that we have affected the natural course of things, such increases in species' populations may be regarded as 'artificial'. However, the increase of a species' membership is not artificial in the same way as 'artificial' rarity is. In the case of the latter, the species are out of tune with their evolved adaptations, they are losing their viability, heading towards extinction. However, with 'artificial commonness' the species' evolved adaptations have come into play and they have maintained their viability. They have merely extended their niche, and there is nothing to suggest that a species' actual niche is the same thing as its realised niche (see Brennan, 1988, p. 49).

Yet there is a further point. If a rare species has become common it must only have been rare in a 'weak' sense of 'naturally rare'. That is, the species must be one that was located in rarely found pockets of habitat, but was not rare within these habitats. Consequently, by increasing the species' habitat, the species' membership increases too. And because it is to be found in a reasonable density within this habitat type, an increase in its habitat will lead to the species losing its rarity and possibly even becoming common. We might call this kind of natural rarity 'clustered' rarity. An example of such a species would be the aforementioned high brown fritillary butterfly that depends upon limestone grassland habitat, or the Red Mangrove that depends upon mangrove swamp habitat (Begon *et al*, 1990; p. 595). Part of the character of species that are 'naturally' rare in a 'clustered' sense is the potential to move out of this state. And I have shown that 'commonness' is no less valuable a state to conservation than 'rarity'. It would therefore seem reasonable to conclude that there is no need for the conservationist to avoid action that causes a species naturally rare in the 'weak' sense, to increase in number. Clearly, then, there is no onus upon the conservationist to maintain natural rarity of this kind. Yet my point may be stronger than this. Where it is the case that the increased membership of species rare in the weak sense is occurring quite naturally, deliberately holding this change back – keeping a species rare – would be artificial in a way that does not sit comfortably with conservation's values. It would not merely be that there is no conservation gain in maintaining such rarity (for different but equal value could be gained from an alternative state such as abundance); there is reason to suppose that there may actually be a conservation *loss* if species evolution and natural development are conservation values.

If certain species are 'naturally rare' in a 'weak' sense, then clearly, others must be 'naturally rare' in a 'strong' sense. These species might be described as those that are rare, not only because the habitat they are located in is rare, but because within their particular habitats they are widely scattered. This kind of natural rarity could be described as 'sparse' rarity. One example of this rarity is the aforementioned *Amorphophallus titanum*. Even if the Sumatran rainforest expanded to cover the entire planet, this species would remain rare, since its increase would be in proportion to the increase of all other forest species. The only way that a species such as this could ever be common would be if all the presently common species in its habitat become very rare or extinct. Even then, the gaps between the members of the rare species would almost certainly be filled by other species adapted to being common, before the rare species had time to become readapted. However, if a species is successfully adapted to its rarity, it seems unlikely that there would be any pressure on it, in evolutionary terms, to become common. If, then, natural rarity in its strong sense is so very robust, so clearly inflexible, then the question of what impact the conservationist ought – or ought not – to have upon such rarity is moot. If conservationists cannot alter the rarity of a species naturally rare in the strong sense, there is no need to worry about the merit – or otherwise – of changes being effected.

Conclusion

I began this chapter by noting that conservationists regularly prioritise rare species over common or abundant species. I wondered whether this indicated that rare species, as a result of their rarity, are more valuable to conservationists than non-rare species. What I have found is that in the majority of cases ('artificial') rarity does not denote a value at all and merely acts as a pointer to the existing value of the species. Although there is good reason to prioritise such species this does not mean they are more valuable than other species; prioritisation is merely pragmatic. Where a species is 'naturally' rare, I found rarity to be neither an amplifier nor a source of value, but part of its existing value. Such rarity is valued only if the species is valued first. However, non-rare species have values that are part of their make-up too, and so there is no reason to consider 'naturally' rare species as more valuable than common species. I have also argued that rarity, as an experience is a source of value.[1] Nevertheless, here, too, it does not follow that rarity is necessarily a more valuable experience than others that relate to species – such as abundance. Although conservation requires that rarity as a *dimension* continues, there is no presumption to preserve each instance of ('natural') rarity. There is certainly no presumption to do so at a cost to any individual members of other species. Indeed, what I have discovered suggests that the implicit association between conservation and rare species does not run as wide – or as deep – as might be supposed. It seems, in fact, that the idea of the importance to conservation of rare species is something of a red herring.

As I noted earlier, insofar as species are 'artificially' rare – that is, endangered – there are good conservation reasons to give them special attention. However, this simply means they require help in a way that non-rare species do not. It does not mean that these species are more valuable than other species. And, with respect to the rarity itself, it is, of course, of no value to conservation. Indeed it is of negative value.

Alternatively, species may be 'naturally' rare. Here, however, whereas we can claim that rare species are of value to conservation, we cannot argue that they ought to be prioritised over non-rare species. If the species is 'naturally' rare in the 'weak' sense ('clustered rarity'), although such rarity is valuable, so too are alternative states of being (such as commonness). Consequently, prioritisation of the rare species cannot be justified. If the species is 'naturally' rare in the 'strong' sense ('sparse rarity'), it still follows that there is no justification for its prioritisation over other species. However, since it is not possible for the conservationist to intervene in such a way that the species becomes common, there is no need to worry about the merits (or otherwise) of such action. Either way, then, there is no reason to suppose

---

1 Recalling O'Neill's claim that 'rarity as such cannot confer value on an object' we must concede that although his argument is faulty, and although we have seen that rarity can, in fact, be a source of value, there is a weaker sense in which his claim is true. It is true if we take 'as such' to mean 'if it ever does then it must always do so'. That is, it is clear that rarity does not of necessity confer value on an object; rarity is not a sufficient condition for valuing.

that conservation has special links with, or ought to have special concern for rare species *per se*; they are of no greater value to conservation than non-rare species. With the implicit assumption that rare species are of particular value to conservation laid to rest, it should be recognised that non-rare species – and perhaps most strikingly common species – are important conservation players in their own right. The value of 'the common' – both in the sense of the abundant as well as the every-day – should be acknowledged.

Chapter 3
Nativity

Introduction

There can be little disagreement that the common aim of conservationists everywhere is the long-term safeguarding of the natural environment. This being so, the very minimum requirement of conservation is that nature remain recognisably nature. From this point onwards, opinions concerning the role of conservation begin to diverge, not least because there is no set standard with regard to what should count as properly representative of nature. Nevertheless, there is a powerful body of opinion which holds that conservation ought to concern itself only with species and other entities that have a 'natural history' in a place and not with species that have arrived there by means of human intervention, intentional or otherwise. This idea is expressed as the belief that native (or indigenous) but not introduced (or alien) species are worthy of being conserved. As noted by Peretti (1998, pp. 183–184), '[t]he protection and restoration of native species is one of the major foci of conservationists' attempts to protect biodiversity. Biological nativism pervades the environmental movement'. Peretti mentions, for example, the proliferation of native plant societies in the United States 'encouraging the exclusive use of indigenous plants in urban gardens'. A British equivalent is the organisation 'Flora for Fauna' which encourages the public to plant native flora in their gardens in order to support native animals and birds and to this end, provides lists of native plants for each postcode area in Britain. In the same vein, Peretti notes that '[s]ince 1963, the United States national parks have attempted to follow the Leopold Report's directive to re-create original ecological conditions – including the restoration and protection of native species' (p. 184).

The pro-native sentiment has two modes of expression. In one – the weak version – it is held that only native species ought to be actively conserved. Such thinking permeates official policy-making. For instance, in Britain, the native red squirrel but not the introduced grey squirrel is given conservation status. And in the USA, a peregrine reintroduction programme was halted when it became known that the reintroductions were of subspecies from Scotland and Spain (De Blieu, 1993). In the other – the strong version – it is held that non-native species ought to be actively exterminated or at least controlled. For example, for a number of years now a campaign has been mounted by conservation bodies (including the Royal Society for the Protection of Birds (RSPB)) to rid Britain of the introduced ruddy duck. It is claimed that in its springtime forays to Spain it hybridises with the increasingly rare white-headed duck, thereby threatening the continued existence of that species (Anon, 1995 b). Pilot schemes took place in 1999 (see Lawson, 1998; p. 40), and in 2003 the Government's Department for Environment, Food and Rural Affairs published the findings from research into 'the feasibility, costs and access requirements necessary to reduce the UK ruddy duck population to less than 175 individuals in 10 years' (defra website). Similarly, efforts have been made to eliminate the grey squirrel because of the damage it causes to trees and the belief that it is ousting the red squirrel. And there have been various crusades against the introduced American mink both because it is thought to compete with the native otter and because it predates the native water vole. Perhaps, though, the most well known hated introduction is the plant *Rhododendron ponticum*, an introduction from Portugal and Spain that conservationists, almost without exception, seek to remove and destroy (see Gritten, 1998; p. 39). Japanese knotweed may, however, soon outstrip the Rhododendron in both invasiveness and vilification. Other introduced species are campaigned against for less obvious reasons. For example, some conservationists advocate the elimination of the small numbers of (harmless, but introduced) snow geese and mandarin ducks in Britain, even though the latter are endangered in their native China and Japan.

My question, then, is this: is the conservation importance attributed to nativity justified? Before I can answer this question, I must first ascertain whether nativeness, as it is currently understood, is a coherent concept, and if not, whether there might be a more appropriate and meaningful way of understanding it.

Is 'nativity' a coherent concept?

The argument that native but not introduced species are worthy of conservation rests upon the assumption that it is possible to clearly identify native species. This, in turn, rests upon the assumption that native and alien species can be distinguished from one another. Yet there are serious difficulties in attempting to do so. One reason for this is that we have insufficient knowledge of the origins of many species. Yalden (1982) argues, for instance, that there are many gaps in our knowledge about native species, particularly for some of the islands such as Orkney, Rum and even Ireland. On the mainland, too, our knowledge is far from complete, even with respect to the history of the rabbit, a species that is almost universally held to have been introduced to Britain by the Normans. Yalden raises the possibility that it might have been *re-introduced* by the Normans, thus implying that it may be an indigenous species (p. 27). Posing the question of how scientists distinguish between native and alien species, Peretti notes that '[t]hey are usually forced to rely on partial natural history records'. He goes on to say that the documented figure of 984 South African alien species is misleading 'because in 104 of these cases, the origin of the alien species is in doubt'. In other words, 'in more than 10% of these cases, the "alien" species might actually be a native'. It seems that '[i]f the natural history record is incomplete, there is no reliable ecological or biological method that can distinguish between aliens and natives' (p. 185). In the same vein, Rackham (1987) writes that:

> [t]he limits of the native British flora are not easy to define. Pollen analysis proves many plants native, in that they are known to have been here since before civilized men, but the converse is more difficult to prove. Plants that produce little pollen or grow in dry places are unlikely to have a pollen record even if they were present (p. 52).

The gaps in our knowledge generate much confusion. For example, the *Field Guide to the Animals of Britain* describes the roe deer as an introduced species (p. 305). Meanwhile Yalden, in his article 'When did the mammal fauna of the British Isles arrive?' describes the roe deer as a native species (p. 21). The reason for the confusion is that, for some, such as Yalden, a species is considered native if it was present before Britain was cut off by rising sea levels about nine thousand years ago. For others (including the author/s of the *Field Guide*) the more recent history of a species is relevant to its status. It seems that although common in medieval times, the roe deer is thought to have become extinct in England and Wales by 1730. It was reintroduced to Dorset in 1800 and East Anglia in 1884. There were claims, though, of a few surviving on in Northern England (Yalden quoting Corbet 1974; p. 41). Yet the same book that classes the roe deer as introduced, then goes on to describe the red squirrel as native. The confusion deepens, as the history of these two species appears to be very similar. The current status of the red squirrel derives (if not entirely then almost so) from a series of reintroductions after the species became extinct in Scotland and Ireland (Yalden, p. 41). So the confusion over the status of 'native' is not merely a result of insufficient historical data, it is also generated by lack of agreement about the meaning of the term.

The practical difficulty of determining species' origins is substantial, but it is clearly not the only problem with the identification of native species. For one thing, as Peretti observes 'it is unclear how long a species needs to be established in a location before it is considered native' (p. 185). As noted above, some sources maintain that, for land-based animals, nativeness is conferred on those species that were present before Britain was cut off by rising sea levels. The importance of the historical element is less clear with regard to the native status of non land-based animals such as seals, bats and birds. They are usually deemed to be native if, as the *Reader's Digest Field Guide* suggests, they breed here (p. 303). However, there is no indication of how long ago breeding must first have occurred. As Peretti asks, '[i]s a species "naturalised" in 100 years, 1,000 years or 10,000 years?'. The distinctions between native and non-native species are, indeed, 'arbitrary and unscientific' (p. 185). We must wonder, then, whether to count as native, species – such as the collared dove – that have colonised Britain of their own accord since its separation as an island. The collared dove, once 'an obscure creature of Asia and south-east Europe' began spreading 'northwestwards across Europe at about 50 miles a year, reaching France in 1950 and Norfolk in 1955' (Rackham, 1986; p. 52). It is now widespread and familiar throughout Britain. And further from home, Livingston (1994) mentions that '[t]he cattle egret immigrated to South and North America from Africa apparently under its own steam' (p. 54). There are also examples of species recolonising after having become locally extinct, such as the avocet (Scott, 1989) or the roe deer. Some species colonised more recently than nine thousand years ago but have since disappeared, such as the pond tortoise that arrived in East Anglia during a warm spell five thousand years ago. It is not at all clear whether any or all of these species ought to be considered native. This leads to the conclusion that, as Peretti suggests, 'the study of biological invasion does not rest on a rigorous scientific foundation' (p. 185).

The problem is that the definition of native lacks a clear cut off point. Indeed, there seems to be no meaningful way of determining one. In addition to 'recent arrival' the primary reason that species are denied native status (and therefore conservation value) is if they have had a helping hand from humans (and often, of course, the two go together). For a species to have native status, it must have arrived in a place *under its own steam*. But what is to count as 'arriving under its own steam'? Even a species – such as the collared dove – that colonises a place of its own accord (or 'naturally') often does so as a result of changes (either in the existing habitat or in the new-found habitat) that have been brought about by humans. Indeed, their arrival may be directly attributed to such changes. Have these species, then, had a helping hand? Is there any way of distinguishing between aided and unaided arrival? What might the reasons be for distinguishing between, on the one hand, the accidental stowing away on a ship and, on the other hand, pollution that leads to an increase in temperature and the spread northwards of less hardy species?

Some may argue that part of the very charm of the natural world is the role of chance. We value the way nature has evolved not merely despite its apparent haphazardness but because of it. But this merely presents a whole new set of difficulties surrounding how to define 'chance'. We cannot exclude all situations involving humans for this would be to argue that chance does not operate in a world with humans. And while pollution-generated temperature rise may appear less affected by chance than the accidental stowing away on a ship, chance could nevertheless play an important part in the former scenario too. For example, where the pollution is produced, the type of pollution, or the way species are affected by the pollution. It is not at all clear that the notion of 'chance' can help us to distinguish between aided and unaided species arrival.

Another question is whether human introductions should be differently evaluated from introductions brought about by other species. Rackham, for instance, mentions the introduction of the (now naturalized) rare cress *Rorippa islandica* that 'grows only on the winter feeding grounds of geese and may have been brought on their muddy feet as they migrate from Iceland' (p. 53). These introductions seem no different in kind from species unintentionally transported by us, such as those that hitch a ride on our many forms of transport. To be consistent, we would have to either include or exclude both from the category 'native'. To consider only those species introduced by humans as alien would be arbitrary in the extreme.

If attempts to distinguish native and alien species on the basis of human involvement are fruitless, attempts to do so by insisting upon the nine thousand year ago cut off point as critical, fare no better. It is difficult to see what the relevance of this historical event could be to the classification of 'native'. Even if it were reasonable to count as native the species present in Britain when the landmasses parted, it seems less reasonable to discount those species that just, as it were, missed the boat. The melting of the ice sheets that resulted in rapidly rising sea levels came about because of the warming of the climate, and this warming was responsible for the movement of species from the south to the north. So those species that were in Britain when it was cut off from the rest of Europe were very much determined by chance. Species were not separated in any meaningful sense. For instance, 'several animal species, such as the tree frog, salamander and garden dormouse reached northern France only after Britain had been cut off by the sea'; consequently they 'never became native to Britain' (*Reader's Digest*, p. 303). The species on one side may just as easily have been on the other. The cut-off point that determines nativeness is quite arbitrary. So, even if we can talk of native species in a *descriptive* sense, it does not seem acceptable to do so in a *prescriptive* sense. Yet those who advocate the conservation of native but not introduced species do just this.

The defining of native species using the shift in landmasses as a critical factor is further flawed in that the cut-off point is not a relevant criterion for the definition of nativeness *per se.* Countries without a history of dramatically appearing ocean boundaries and the cessation of natural dispersal cannot define native land animals in this way. The cutting off of Britain by the sea does not, then, have much credibility as a criterion for defining native species. Yet without this criterion, it is difficult to see how native land animals could be defined any differently from non-land-based native species such as birds. That is, it is difficult to see why they would not, too, be deemed native by virtue of their breeding in Britain. There would be no reason to exclude even very recently arrived animals, provided they bred in this country. Yet even the requirement of breeding here has to be questioned. Any book of British birds will include those that over-winter – but do not breed – in the British Isles. Even if it is fair to expect different criteria for migratory species, it still remains true that the definition of a native species is extremely difficult to arrive at. Combining these findings with those above concerning native species, it becomes clear that the definition of native is very loose indeed. The difference between our current understanding of the terms 'native species' and 'introduced species' is extremely blurred.

To suppose that native species can be rigidly defined either in terms of a particular historical cut-off point and/or in terms of unaided arrival is muddle-headed. It also leads to conclusions that are counter-intuitive. Yalden has good evidence that many of our islands gained their resident species of animals by way of humans. We would have to say, then, that a number of islands such as Rum and Orkney have few or no native species of land animal, even though we might think of, say, field voles as belonging in the Outer Hebrides or on such western islands as Muck or Eigg. The Soay sheep, though arriving on Soay at least three thousand years ago with Stone Age farmers cannot be classed as native, despite any feelings to the contrary. Yalden writes that although Ireland and the Isle of Man probably acquired their native species via land bridges, '[o]ther islands, Shetland, Orkney, St. Kilda, Outer Hebrides, have never been connected to any mainland, and their terrestrial mammal faunas probably result entirely from human introduction (pp. 51–52)'.

Not only must we accept – on the current understanding of the term 'native' – that some of our islands may be devoid of native species, we must also accept that some of the species we think of as native to the British mainland are discounted. The house-mouse and the little owl are two such animal examples, the former being known to have lived here in the Iron Age (having probably been accidentally introduced) and the latter a nineteenth century introduction from the continent. Both species are widespread, integrated and well-known members of the British fauna. It would probably surprise many that they cannot be classed as native. Similar cases can be made for the pheasant and the fallow deer. Many, too, would undoubtedly be surprised to learn that the Sweet Chestnut tree, *Castanea satira,* is not a native member of Britain's flora. Indeed, as Rackham points out, '[u]ntil the invention of pollen analysis we did not know that it was not native ... [e]arly writers such as

Evelyn treated chestnut as indigenous' (p. 54). A whole range of trees we might imagine to be native are probably – or certainly – introduced, such as common pear, sycamore, common walnut, white poplar and cedar. For most people, nativity is not easily or readily defined by the rigid criteria described above.

I have shown there to be many difficulties with the rigid definition of native that conservationists attempt to adhere to. There are good reasons, then, to abandon a definition of native species based upon arrival either at a specific historical time or unaided by humans. Does this mean, though, that there is no coherent understanding of the concept of 'native species' and, by implication, that the conservation importance attributed to nativity cannot be justified? Is it the case (leaving aside the issue of rarity) that the conservationist ought to consider all species to be found in an area as having equal conservation value? Must we assume, for example, that the rabbit is as worthy of conservation in Australia as the koala bear?

Are native species of particular conservation value?

Although I argue that the above formal understanding of nativity should be rejected, I do not propose that the idea of native species should be abandoned altogether. There is reason to believe that nativeness is a coherent concept. There is also reason to believe that it has significance and value to conservation practice. Though nativity cannot be defined in a rigid, or formal sense, the idea of native species carries with it the informal but intuitively important sense of *belonging*. When we think of native species, it is likely that it is this idea above any other that comes to mind. The notion of belonging is inherent in the concept of native. It seems to capture, too, what is worthy and intuitively important about nativity that an insistence upon absolute boundaries and dogmatic prescription misses. Insofar as belonging is an intuitive, non-empirical concept, it is more appropriate in the context of our relationship with the natural world, a relationship that is dynamic and ever changing. Nevertheless, though the idea of belonging may defy precise definition, we still need to know what kind of thing it is. In fact, there are three related questions. Firstly, what do we mean when we say that something belongs? Secondly, why is belonging significant or valuable to conservation? And thirdly, what follows from the knowledge that an organism belongs – or does not belong – in a particular habitat?

To say that a species belongs might be explained by saying that it has a niche (defined by Paul Colinvaux as the 'role, function, or place of [an] organism' (quoted in Livingston, 1994; p. 66). However, some authors have pointed out, as Livingston does, that '[t]here is something tautological about the concept of niche' . The reason for this claim is that 'a niche exists only when it is occupied' (Livingston, p. 67). In the words of Donald Worster, says Livingston, 'the niche is the species and the species is the niche'. That is, '[i]f it is there, ergo, it has found a niche. It has won the competition for a place' (Worster, 1985, p. 299). Perhaps, then, the concept of niche is unhelpful as a way of articulating a sense of belonging or of clarifying what is meant by nativity.

It may be, though, that both 'niche' and 'belonging' can be understood in a way that is not tautological. Livingston points the way. He reluctantly asks, for example, whether we must infer that the feral goat 'on its denuded Galapagos hillside ... has a niche after all [for] [i]t is plain that the goat is there, and that it *does* things'. He goes on to say that '[i]t even interacts with other species – negative, to be sure, but it clearly has a "role" in the community, no matter how repellent that role may appear to us' (P. 67). Nevertheless, Livingston feels that 'few if any ecologists ... would see any feral domesticate as "belonging" in the sense of contributory integration – of properly occupying either a spatial or a functional space in any natural community' (p. 68). Livingston sees it as a problem, then, that the competition-orientated ecologist, dictated to by market-place analysis,[1] must consider the feral goat as belonging in the above sense. His solution is to argue that 'where the way has actually been prepared for the exotic domesticate [competition] would not be a factor'. Non-human feral domesticates, he believes, 'have a much better chance of "succeeding" if the way has been paved for them by some measure of prior human settlement'. This is not, he says, a 'competitive' situation. So for Livingston, domesticates, though surviving in nature as ferals, cannot be said to have a niche because they have not secured their place through the proper competitive process.

Livingston's argument that feral animals (domesticates that have returned to the wild) *per se* cannot be considered to have a niche or to belong is not convincing. After all, they were once wild animals, and as Livingston himself admits in his discussion of the Galapagos goats, they can pretty much regain their wildness:

> Goats are fast multipliers, and they have been here for more than three centuries. Not having been subject to artificial breeding selection for all that time, they have somehow ... not only "reverted" to keen alertness but also regained many of the other behavioural and physical attributes of the wild ... It is clear that they have shed many of the earmarks of domestication, perhaps most of them (pp. 46–47).

Livingston's claim that feral animals cannot belong is much too sweeping. His argument is skewed towards showing that the human (and consequently all human by-products, as it were) are placeless aliens (he calls them 'domesticates')[2] in the natural world. The human mark upon feral animals does not seem sufficiently recent, profound, or lasting to label such animals as ecologically placeless beings, beings that *cannot* belong. Nor is Livingston's claim that feral animals do not take part in the competitive process convincing; he fails to show in what way they are exempt. Furthermore, it would be very difficult to draw a line between what does and does not count as being aided by humans (see Chapter Four for more about this). However, the main problem with this particular claim is that Livingston tries to show that feral animals do not belong by way of the ecologist's economic imagery and language – namely, 'competition'. The idea of belonging cannot be captured by this kind of explanation; it does not lend itself to marketplace analysis. Rather, it is a term that is engendered by – and engenders – emotions or feelings such as comfort,

---

1 Livingston (p. 65) is referring to the ecological maxim known as the 'competitive exclusion principle' that is often translated as 'competitors cannot co-exist' which, he suggests, is not a biological principle but an economic one, and one that fits neatly into the economic structure of the marketplace.

2 For clarification of Livingston's term, see Chapter Four.

security and contentment in familiarity; pride in local distinctiveness; protectiveness in local 'ownership'; love of 'one's own', and so on. Livingston's mistake in trying to argue for the notion of belonging as part of the way in which niche should be understood is to ignore this human element of the concept. Indeed, what I suggest is that this, too, is conservation's mistake. As Ian McLean, co-ordinator of the UK's Joint Nature Conservation Committee acknowledges, conservation is not only about scientific measurements, but 'also about people's perceptions and interests ... Conservation is a human-centred activity, and you can't do it in a scientific vacuum' (quoted by Lawson, 1998; p. 40). This is never more relevant than with the issue of native species. While scientists adhere to rigid pseudo-scientific determinants of nativeness, the public are concerned, instead, with the way they *feel* about the wildlife around them. Lawson quotes Richard Mabey as saying:

> One could generously say that part of the anti-alien passion at the moment may spring from the same sort of passion that champions indigenous cheeses and beers ... But if you test that at local level, people are not in any way concerned about the nativeness of wild creatures. Their passion is more to do with a long degree of association with those creatures (p. 40).

And the same goes for plant species, about which Mabey is said to remark: 'many of the species most loathed by botanists are loved by local communities'. Mabey suggests that people are not concerned about formal nativeness. However, given that the idea of belonging is, I have argued, to be a more appropriate understanding of nativeness than the existing definition, such local opinion could be interpreted as a pro-native stance. Nativeness, then, is more appropriately to do with securing a place in the heart of the local community. Some conservation bodies do recognise this. Lisa Kerslake, policy co-ordinator on the red squirrel for the Northumberland Wildlife Trust is quoted as saying that 'we have no intention of calling for culls where grey squirrels have become a popular part of local culture' (Lawson, p. 40). If, then, conservationists wish to involve and enthuse ordinary people in nature conservation, they would do well to take on board such peoples' understanding of what it means for a species to be valued or to belong.

Certainly, understanding nativeness as belonging goes some way towards avoiding a particularly unsavoury association that its current, formal definition engenders. This is the link with right-wing nationalism and racism. (Indeed, it is a more extreme version of the – official – right-wing attitude towards people, for we are classed as native to Britain if we have been born here. Even the Angles and the Saxons – whom the English identify themselves with – were fairly recent arrivals in Britain, appearing on the scene around the fifth century (Morgan, 1984).) The conservation emphasis on native species as the 'good guys' and aliens as the 'bad guys' can be interpreted – by those who are so inclined – as racism, and used to justify racist attitudes towards humans.[3] As Peretti (1998) notes:

> [a]lthough it is impossible to prove an essential link between particular forms of scientific knowledge and the societal context from which they emerged, the purism of biological nativism has historically been associated with fascist and apartheid cultures and governments.

The Nazis, for example, '[i]dealogically, politically and ecologically ... sought to prevent mixing and to purify categories. They attempted to purify nation and nature, by eliminating people and biotas that were supposedly not native' (p. 188). And it is no exaggeration to liken some conservation policy today to Nazi-like operations. The lengths to which conservationists will go to maintain the 'purity' of native species could not be better illustrated than by the following letter. It was sent by naturalist Richard Mabey to *BBC Wildlife* magazine in response to an article in the previous issue. This told of the disappearance of five eggs laid in Orkney from the first ever known union between a female hen harrier and male pallid harrier that had strayed from its Russian homeland. The letter is quoted in full:

> Thanks to an impeccable, but confidential source, I think I can shed some light on the fate of the hybrid hen/pallid harrier eggs that disappeared from Orkney this Spring (Discoveries, September).
>
> The thief wasn't a 'hooded crow', but a crack commando team from CANT (The Campaign Against Nasty Translocations), honed to a ruthless efficiency by their recent raids on ruddy duck sites.
>
> Their intervention was ordered by the Foreign Office, who blanched at the consequences of telling President Yeltsin to eradicate his country's pallid harriers because they were showing dangerous signs of becoming aggressive imperialists, bent on hybridising our native harriers out of existence.
>
> Better, they thought, to quietly nip the whole thing in the bud (Mabey, 1995a, p.97).

Not only is the current pseudo-scientific definition of native worrying in its overt associations with racism, but, as the above example illustrates, it also creates tensions within conservation biology. The first is that between the conservation biologists' emphasis of 'the importance and pervasiveness of species migration while maintaining a nativist ideology' (Peretti, p. 185). And the second tension is that between the conservationists' efforts to prevent interbreeding between an alien and a native species and conservation biology's most widely-used definition of a species as 'interbreeding members of a population' (Chapter Six deals with species definition in detail). Difficulties such as these are greatly reduced, if not dismissed if native species are defined, instead, in terms of belonging. This is partly because belonging is a more inclusive understanding of native than the present one. It is also

---

3 I acknowledge that non-belonging can also be used as a reason for exclusion and prejudice. However, as becomes clear shortly, the notion of belonging is a great deal more inclusive and reasonable than the current, rigid, exclusive and overly narrow understanding of native.

because the concept of belonging – as loose, dynamic and socially constructed – can never be rigidly prescriptive.

It may be objected that it is all very well to suggest belonging as the new way of understanding nativeness, to argue that people's attitudes towards species are relevant and important to conservation, but what becomes of the conservationist's primary reason for protecting native as opposed to alien species? What becomes of the belief that native species are better fitted to their environment than are alien species? In particular, native species are said to support more life than introduced species, thereby maximising biodiversity, while introduced species are accused of decreasing biodiversity. For example, the introduced plant *Rhododendron ponticum* is said to be 'destroying our native ecology'. Rod Gritten (1998), an ecologist at Snowdonia National Park, goes on to say:

> Like so many invasive aliens, once it has been moved from its own ecosystem, it finds itself with no enemies – no animals or insects to eat it, no pathogenic fungi to control it, and both a climate and soil that favours its growth and spread ... R. ponticum is toxic to mammalian herbivores. Few insects seem to feed on it. Its roots produce herbicidal chemicals that suppress the growth of other nearby vegetation. And it casts such a dense shade at close-canopy stage that nothing can live beneath it. Where rhododendron takes a hold, nothing else survives ... The unchecked spread of this plant is an ecological and landscape disaster (p. 39).

If people's feelings towards wild species were to be prioritised, then presumably, if *Rhododendron ponticum* became a much-loved aspect of the local ecology, there would be no reason to consider it of less conservation value than the array of plants that would grow in its absence.

If people's preferences *per se* were all that counted, then the above worry would be justified. However, we are not talking about preferences in some kind of vacuum, but rather, preferences hitched to the notion of conservation. I argued, earlier, that when we say we value nature, we mean that we value it in something like its current, recognisable form. The need for conservation has arisen as a means of securing this value threatened by our own activities. So it is clear that conservation is a human enterprise, with human values. However, it is also clear that conservation must be opposed to (or at least not actively in favour of) anything that threatens the kind of form nature currently takes. There is a sense, then, in which conservation supports some kind of equilibrium in nature; too much or too rapid change is threatening.

To argue for the importance of equilibrium of some kind is not to take the view that nature is a stable system in which a balance is always maintained. This unsatisfactory and mistaken view of nature is one that the current definition of native is associated with: native species are seen to be maintaining the balance while alien species disrupt it. Peretti (1998) discusses the recent academic rejection of the balance-of-nature perspective. He cites a number of authors (such as Worster, Hengeveld and Soule) who emphasise the instability, movement, haphazardness – even chaotic character – of nature. Conservation and population biologists, he writes, 'tend to view species migration as natural and normal'. Indeed, the former 'emphasise the importance of biodiversity and have identified free species migration as a central element in preventing species extinctions'. As Peretti observes, '[t]his theoretical shift in certain biological disciplines challenges most of the previous work on biological invasion' in which aliens are seen as damaging because they 'disturb the balance of the ecosystem' (p. 187).

There is much recent evidence against the ecosystem balance-of-nature paradigm, providing further ammunition against the formal notion of native to which it is wedded. Nevertheless, it still remains true that since conservation seeks to safeguard nature in something like its current form, change of too rapid or disruptive a kind must be detrimental to the aims of conservation. That is, there is a notion of equilibrium that is relevant to conservation, but it is a human-centred rather than eco-centred notion.[4] An important element of this equilibrium is that it can be recognised to pertain within the human time-scale. It may well be that in terms of an evolutionary time-scale certain apparent disruptions of the natural world would ultimately be viewed positively. However, neither we – nor conservation – can operate within an evolutionary time-scale. What is important to us is that changes in nature will not ultimately lead to loss of the natural world as we know and cherish it. For this reason, we value more those species that perpetuate this notion of equilibrium. These are necessarily those species that we can see as being seamlessly integrated into the natural community, or, in Livingston's terminology, those species that can be said to have a niche in terms of 'contributory integration'. That is, we value more those species that form positive associations with the other resident species in terms of food, shelter, pollination, parasite control and such like. These species are very much a part of – and help to maintain – the nature we know and cherish. So the notion of belonging (our new meaning of native) is not wedded to any formal idea of a balance-of-nature (as nativity currently is). Yet it cannot be entirely divorced from a human sense of equilibrium in nature if it is to have meaning in a conservation context. Any talk of belonging, then, even in Livingston's ecological sense of 'properly occupying either a spatial or a functional space in any natural community' (p. 68), has to acknowledge the human element. Ecology cannot tell us, for example, what constitutes *properly* occupying a space.

The conservation notion of belonging entails, then, the idea of equilibrium from a human perspective. This notion of equilibrium is really one of 'fittedness'. When we ask whether an organism belongs we are asking whether its presence 'fits' with its environment. We are not asking whether the newcomer is ecologically fitted to its environment in the sense of whether or not it is *successful*, but rather, whether it fits in with its environment. For instance, does the newly arrived species allow the resident species to adapt to it? If there is an obvious sense in which adaptation is not occurring (rapid extinctions, for example) then our sense of equilibrium is upset. Such a species could not, then, be said either to belong or to have conservation value. Where there is, what we might call 'seamless integration', where the new species fits in without causing too much disruption to our anthropocentric notion of the natural equilibrium, conservation value is retained.

---

4 This notion of an anthropocentric equilibrium might be thought to allow for too much change of the wrong kind. It has to be remembered, though, that change of any kind must be tempered by the need to maintain nature in 'something like its current form'. This restriction upon change is further articulated in Chapter Four.

In order to illustrate the point about 'seamless integration' I can do no better than turn to Aldo Leopold's (1966) *Sand County Almanac* in which he bemoans the extirpation of wolves from numerous North American states. In this case, the idea of belonging is highlighted by the loss rather than the gain of a species, but the point is the same. Leopold writes:

> I have watched the face of many a newly wolfless mountain, and seen the south-facing slopes wrinkle with a maze of new deer trails. I have seen every edible bush and seedling browsed, first to anaemic desuetude, and then to death. I have seen every edible tree defoliated to the height of a saddle-horn. Such a mountain looks as if someone had given God a new pruning shears, and forbidden Him all other exercise. In the end the starved bones of the hoped-for deer herd, dead of its own too-much, bleach with the bones of the dead sage, or molder under the high-lined junipers (pp. 139–140).

It is clear that, for Leopold, the wolf belongs on the mountain, for without it, there is no sense of equilibrium; the change is too rapid, of too great a magnitude, for him to feel comfortable with it. Leopold's use of language and imagery indicate, quite clearly, that this notion of belonging is not merely an ecological, empirical concept, but also a personal, emotive, passionate and poetic one that centres on the idea of loss. It is not that the former situation could never prevail again. Rather, it is the fact that it may not be possible in a human life-time: 'while a buck pulled down by wolves can be replaced in two or three years, a range pulled down by too many deer may fail of replacement in as many decades'. Nor is the fact of change in itself significant. Sheep and deer have similarly denuded the Scottish Highlands, yet to many, the new relationship between moorland and deer, and the peculiar beauty of the close-cropped mountains are part of what is cherished about this area. These things have become part of the character of the place; they have gained cultural significance. The point is that our idea of what does or does not belong is not static; it shifts, or evolves. It is not, though, an arbitrary notion. It is rooted, as I have illustrated, in whether the incoming organism can be adapted to by the resident species in a way that satisfies our anthropocentric idea of 'equilibrium' or 'contributory integration'. So, to belong is to be ecologically integrated in a way that is both meaningful and relevant to us. It is to have attained a place in the hearts and cultural history of the local people. It is to be accepted.

The notion of belonging, then, is primarily socially, culturally, or personally-based. It is about a species being seen to be part of a local ecological community. This being so, two points require clarification. Firstly, why, exactly, should belonging *matter* to conservation? And secondly, what constraints – if any – might there be upon belonging? In other words, are there any species that could be said *not* to belong?

There are two reasons for claiming that belonging should matter to conservation. I have already noted that conservation is a human enterprise. That is to say, conservationists decide upon the way that nature ought to be and make decisions with respect to how this is to be achieved. No scientific data or formula can tell us how nature *ought* to be. I have revealed that one of the hooks on which conservationists try to hang their decision-making – the idea that native species are best – is problematic. The accepted definition of native does not stand up to scrutiny; nativeness cannot be easily defined. Nevertheless, something of value has been extracted from the idea of nativity – the sense of belonging to an ecological community. This is a notion that is already close to the hearts of many ordinary people. A species is often seen to be of value and worthy of protection when it is a familiar part of the ecological community. It may be fair to argue that conservation ought not to be the exclusive domain of an elite group of scientists, and that it ought to take account, too, of the opinion of ordinary people. Whether this is so or not, it is certainly reasonable to argue that conservation will prove an uphill struggle without the approval, enthusiasm and energy of ordinary people. For this reason alone, then, whether or not a species is seen to belong ought to be taken seriously by conservationists.

However, I am not suggesting that conservation should consider all and any ordinary people's views on the matter of which species are worthy of conserving. People's preferences, as such, are not necessarily of value. The notion of belonging encapsulates rather more than mere preferences. Recalling that biodiversity is a fundamental conservation value, it becomes clear that the notion of belonging is worthy of conservation note, because those species that belong are more likely to be those that are best for biodiversity. That is, those species that have been around a long time – and have had time to become seamlessly integrated – will, by definition, belong. Species that are well integrated will be likely to be those that support numerous other species and are therefore good for biodiversity. What *cannot* be said, however, is that there is a *necessary* connection between belonging and biodiversity value. Given that the current definition of native has been revealed to be so problematic, any claim that a necessary connection of this kind could be made between old version native and biodiversity value cannot be upheld either. There may, then, be instances where a species is said to belong despite having little biodiversity value. For the most part, though, there will be a strong correlation between the two.

Having shown why belonging matters to conservation, I now need to ask whether there are any constraints upon species being said to belong. For example, might the conservationists' scourge *Rhododendron ponticum* be thought to belong, for it is a very familiar species in Britain today, and its flowers brighten many a park? Bearing in mind that belonging is not merely a preference, but an ecologically informed preference, it seems unlikely that rhododendron could be a candidate, given its poor integration with other species. Remember that 'no animals or insects ... eat it ... few insects feed on it ... and it is toxic to mammalian herbivores' (Gritten, 1998, p. 39). But if we can say this about one traditional enemy of the conservationist we may not be able to say the same about other traditional enemies. Consider, for example, the feral goats living on the Galapagos islands.

Livingston (1994, p. 47) echos Leopold's thoughts in his description of the Galapagos goats as placeless, a placelessness that 'rings from the desertified hillsides, gleams from the whitened brittle carapaces of long-dead tortoises'. However, what is interesting about this case is that it is not a foregone conclusion that, on the basis of the foregoing argument, the goats must be considered not to belong here. For one thing, they have been present, as Livingston notes, for more than three centuries. It is therefore quite feasible that they are now part of the character of the place. They may be significant, at least to local people, perhaps also to some naturalists, in the same way as the Australian dingo – an introduced feral species – is of interest to naturalists. But for another, the goats are part of a very complex set of inter-relationships both with wild and other feral animals. Their removal (even if possible, which is unlikely) would probably result in greater loss of 'balance' than their presence might be causing. It is worth reproducing, in full, part of a 1983 report quoted by Livingston:

> This is hardly a simple case of bad (recently arrived) guys vs. good (older resident) guys. For one thing, there are complex interrelationships among the exotic species themselves. The dogs, for instance, attack feral cattle and help keep cat populations in check. Any strategy aimed at totally wiping out feral dogs would have to take these kinds of facts into account.
>
> Other facts complicate any attempt at an overall battle plan for reducing exotic species. Goats, for example, were formerly selected as the prime target for eradication, until hunters realized that, in the absence of goats, vegetation quickly grew back, making it extremely difficult to spot the wild pigs slated for destruction (p. 47).

The argument, focusing here upon the feral animal as incomer, relates every bit as much to (exotic) wild animals. For example, given the rapid and epic changes the rabbit has wrought on the Australian landscape and wildlife it may be fair to conclude that it does not belong there. Livingston notes that:

> [t]he impact of the rabbit was catastrophic. Country that had supported lush grass was now eaten down until it resembled a bowling green. At times of drought the rabbit died by the millions, but before doing so they ring-barked and killed shrubs and bushes, and ate grass down to its roots (p. 40).

Further, the impact that the rabbit has had on the land has resulted in the disappearance of 'several grassland-adapted marsupials and a number of birds', some of which succumbed to 'the devastation of their habitat by the intruders' (p. 41). Nevertheless, even this case, perhaps the best known of its kind, is not entirely clear-cut. Not all of the species were lost through habitat changes. Some 'succumbed to anti-rabbit poisoning programmes'. Furthermore, some of Australia's predators have come to rely upon the rabbit for their survival such that rabbit extermination poses a threat to them. And finally, it does seem that the rabbit, present in Australia since its 1859 introduction, has become part of the character of the landscape – if not in a cherished sense, then at least in the sense that any species present in such enormous numbers has a certain significance and clearly cannot be ignored. The mere fact that a species' numbers reach plague proportions does not exclude it from the current formal definition of native. Neither need it do so with respect to our new understanding of native as belonging: animal plagues have become part of folklore in the past (the Pied Piper tale being one such example). The rabbit 'plague' may well become part of Australian folklore.

If the case for the rabbit in Australia is hard-won, other cases for introduced species are more readily convincing. The house sparrow and the common starling were introduced to North America in the 1800s. They have both 'blanketed almost the entire settled portion of the continent' (Livingston, p. 39). Yet though they have had some impact upon existing species, '[n]either ... appears to have had truly grievous effect. For destructiveness they have not been in the same league with agriculture and urbanization' (p. 40). Indeed, Livingston observes that '[a]lthough both the sparrow and the starling now number in the hundreds of millions in their new hemisphere, their disruptive interspecies accomplishments pale beside those of the infamous European rabbits in Australia'. Even in cases where so-called destruction of some kind occurs (and for an in-depth discussion of what destruction might mean, see Chapter Four) this does not necessarily rule out the possibility of belonging. Many people in Britain, for example, have taken the introduced grey squirrel to their hearts, even though it causes some damage to trees and appears, in parts, to be out-competing the longer-established red squirrel. On the other hand, ironically enough, the proposed re-introduction to the British Isles of the so-called native beaver met with some opposition on the grounds that these animals will destroy trees and change the landscape through their dam-making (see Blundell, 1998). The notion of belonging is not a straightforward one. In the case of the beaver, its long absence from the British landscape may (at least partially) have severed its place in the hearts of the people. It will be interesting to see whether attitudes begin to change as the beaver finds its feet.

Though the notion of belonging is multi-faceted and may not be easily defined, it offers an understanding of the term native that is meaningful, realistic and in tune with the human-centredness of conservation. The current definition of native that rests upon supposedly empirically defined boundaries must be rejected as meaningless. Its boundaries are arbitrary and the empirical details are often unknown or unknowable. My revised understanding of nativity as belonging is of more relevance to conservation because it succeeds in taking into account the human/cultural perspective. Without this the notion of conservation makes little sense and can gain no purchase in the real world.[5] A distinction between native and non-native species is of relevance to conservation, then, if what we mean by a native species is one that belongs. For a species to belong it fits in, in a way that is meaningful to us in an ecological as well as a cultural sense, both of which are essential components of the conservation idea. This new understanding of native is not prejudiced against some cases of recently arrived species. That is, some species currently regarded as exotic introductions would transfer into the category of 'native species'. The distinction, then, between indigenous and introduced species is no longer relevant. The distinction that matters is between species that belong – which I am calling native species – and species that do not belong – which would best be described as alien species.

Having reworked the concept of nativity, I must now consider the implications for conservation action. Though our understanding of nativity has changed, it is still true to say that native species (those that belong) are of greater conservation value than are alien species. Indeed, it would seem fair to say that native but not alien species are worthy of active conservation. Currently we generally seek only to actively conserve those species that we regard as native in the formal, historical sense. With our new understanding of nativity we might, for example, seek to protect the reindeer. Though reintroduced to Britain in 1952, it is very much a part

---

5 More will be said about this in Chapter Four.

of the character of the Cairngorms in Scotland. Similarly, the little owl, a great favourite amongst bird-watchers, though an introduction, could be afforded protection where needed. The mandarin duck and ring-necked parakeet could be considered worthy of protection too. Conversely, we would do nothing to aid the spread of the recently arrived New Zealand flatworm that is allegedly decimating the resident earthworm, an important improver of our soil. Such a species cannot be regarded as belonging either in an ecological sense (resident species are failing to adapt to it) or in a cultural sense (it has not played a part in local distinctiveness or folklore).

Currently, there are those who argue that non-native species ought to be exterminated. There is, however, no firm basis from which to argue that species that do not belong ought to be exterminated. The reasons for this are two-fold. Firstly, the concept of belonging is not a precise one: it is not *always* possible to say whether something belongs or not. As I – controversially – mooted, it is not absolutely clear that the European rabbit does not now belong in Australia. The concept of belonging (and the updated version of nativeness) is open to a degree of interpretation. And secondly, whether a species belongs or not is not a static affair. A species that does not belong today may do tomorrow. So, while it is reasonable to withhold active conservation measures from a species that is considered not to belong, there does not seem to be sufficient justification for its *elimination* on this basis alone. This is because species elimination is quite final. Any action with such non-reversible consequences ought to be based on very clear category distinctions that are prone to neither interpretation nor change.[6] Certainly, where exotic species are causing the rapid extinction of resident species, they are clearly acting against the interests of conservation that seeks to safeguard nature in something like its current form. For this reason I readily admit that the deliberate introduction of species to new areas (an action that we necessarily cannot know the outcome of) ought to be avoided. However, once the species has arrived, it is not clear that its non-belonging is sufficient justification for its attempted extermination, even if it is having a deleterious effect upon the resident wildlife. Both killing and species extinction (even where this is local as opposed to global) are very serious issues in terms of their finality.[7] It would therefore seem reasonable to require very clear-cut boundaries for those to be categorised as suitable recipients of such action. But such clear-cut boundaries do not exist. Furthermore, there are other reasons that add weight to this thinking. To begin with, the removal of such a species is often impossible and many such attempts have proved costly in terms of money as well as time. These are losses that conservation could well do without. In addition, attempts to exterminate unwanted species very often have a negative effect upon the very

---

6 I acknowledge that when conservationists advocate elimination they are not advocating global elimination. Nevertheless, much localised elimination can be final in the sense that once the species has gone it is not likely to return without help.

7 There is a wealth of literature that argues very strongly for each of these claims. There are several references with respect to the seriousness and finality of species extinction to be found in Chapter One. The seriousness of killing sentient animals is dealt with (amongst many others) by Singer (1991), Regan and Singer (1976) and Regan (1984).

species that removal seeks to safeguard. As Livingston remarks (rather resignedly) in his discussion of the effects of feral animals upon the older resident Galapagos species:

> No medication, it would seem, is without a side effect of some kind. Since prevention is no longer an option in most parts of the world, one cannot but wonder whether the best remedy might by to let the feral disease run its course ... (pp. 47–48).

Perhaps, though, the most worrying problem with species eradication that has implications for conservation is the ethical justification for killing. It is not at all clear that it is morally acceptable to kill members of one species in order to safeguard members of another species. This is particularly true when the unwanted species members are sentient. There is, of course, a long history of philosophers who note the moral importance of sentience, including Jeremy Bentham and, more recently, Peter Singer. The suffering that is caused by such killing may, in itself, be unjustifiable if it is accepted that sentience is morally significant. Yet further, insofar as killing sentient animals is unacceptable to members of the public, such action reflects badly upon, and may seriously damage the image – and the fortunes – of conservation. The public outcry against the planned ruddy duck cull has been significant. Judy Ling Wong observes that:

> [t]he ordinary person loves all fluffy animals and doesn't care whether they are native or not ... People do not want conservationists to kill grey squirrels because they drive out red squirrels. It is "conservation" that wants this, and it projects emotive values onto species to justify the culls (Quoted by Lawson, 1998; p. 40).

Although it is often said critically that the public is only concerned about 'fluffy' or 'cute' animals, there is evidence that concern is wider than this. Concern quite evidently extends to sentient animals outside these categories: frogs, newts, rodents such as water voles and whales, for example. Sentience is part of the reason why members of the public are incensed when conservationists seek to eradicate unwanted introduced species in order to safeguard desirable so-called indigenous species. It is not, however, the full explanation. As I have established, the public values species when they are felt to belong and for this the species does not require sentiency. A species' sentience is an additional reason for the killing of its members to be deemed unacceptable by the public.

Conclusion

In conclusion, then, I argue that native species (as they are characterised here) but not non-native species, are worthy of proactive conservation. There is not, however, any justification – in conservation terms – for the extermination of non-native species. There are insufficiently clear or static boundaries between the categories of native and non-native, a requirement, I argue, that is necessary for action as drastic and final as extermination. Indeed, I suggest that there are good reasons to suppose such action to be harmful to conservation. The moral implications of killing sentient animals offer a powerful argument against the extermination of members of a sentient species in itself (that is, each individual animal that is killed is harmed). But the public hostility that killing engenders is a powerful conservation argument against such action, for effective conservation so often relies upon public goodwill. It would be fair to say that many regard sentiency as an overriding reason against killing and that this includes any supposed conservation benefits. Although other species are often valued greatly despite their lack of (or questionable) sentiency, there may well be cases where their extermination is justifiable, even if not for conservation reasons. It may, for example, be justifiable to eliminate disease organisms or species that destroy crops. However, where such good reasons are to be found, they will be likely to apply, not only to non-native or alien species but to native species too. That is, the extra value conferred on a species by virtue of its belonging would not be likely to afford it much more protection than those species that did not belong. This is so because belonging is such a nebulous term, but also because killing is a serious issue requiring serious reasons. If reasons for killing were sufficiently serious, they would very likely override reasons for protecting something that belonged. So-called native species are significant to conservation, but less so than conservationists would currently have us believe.

PART II
THE CONSERVATIONIST'S ROLE REDEFINED

Chapter 4

Humans and nature: is conservation meaningful?

Introduction

In Part One, I considered three areas of conservation not normally examined, in order to arrive at a clearer understanding of what constitutes good, coherent, conservation practice. I considered, first of all, whether human-caused extinction is different in kind to other forms of extinction in order to discover whether the conservation assumption – that the current extinction event is undesirable – is coherent. That is, I wanted to discover whether today's extinctions are different enough from extinctions of the past to warrant different appraisal. I found that the current extinction event is, in fact, unprecedented. This shows, at least, that the conservation assumption that extinction is undesirable rests upon a coherent foundation.

I considered, too, the assumption that rarity – a familiar focus of conservation policy – ought to have the importance often attributed to it and I discovered that its value, where it does exist, is no greater than the value of other states such as commonness. And finally, I questioned whether native species ought, as is currently accepted, to have greater conservation value than non-native species. What I found was that the current notion of native is untenable, but that a more intuitive, less binding idea of native – namely, the idea of belonging – has some conservation importance.

By questioning these commonly accepted conservation assumptions, I have arrived at a somewhat different idea of where conservation's priorities ought to lie, and therefore of how best to conserve nature. However, a fundamental issue remains, as yet, unexplored. In discussing what constitutes good conservation practice, I make the implicit assumption that it makes sense to talk about human interaction with nature. In doing so, I may seem to assume a dichotomy between humans and nature. In this chapter I explore the relationship between humans and nature with a view to determining whether these assumptions are correct.

PART 1
Humans and nature: a dichotomy or a continuum?

It is commonplace to hear the accusation that by our actions we are interfering with, impacting upon (or at the very least, interacting with) nature. To talk in this way suggests that humans are something *other* than nature. While the majority probably accept this as an unsurprising fact, there are some who argue that humans and human activity are no less natural than those things commonly accepted as features of the natural world.

This not only leads to a rejection of any notion of 'human impact' or 'interference', but for some, to the claim that any human activity is as natural as any other. (For example, this line of argument has been used by developers who seek to replace meadows, woodland and the like with car parks, office blocks and so on). If this were found to be so, then the entire conservation project – based as it is on the importance of 'the natural' – would have to be rejected as meaningless. On the other hand, there are those (such as wilderness proponents) who take the view that *any* human activity compromises 'the natural', implying that humans and nature are utterly different – indeed opposite – entities sharing no common ground. My task, then, is to discover whether either of these extremes is truly representative of the relationship between humans and nature, and if they are unsatisfactory, whether a better representation can be found.

The question about the relationship between humans and nature is tied in with the question about the definition of nature. When we know the answer to one, the answer to the other becomes clearer. For this reason, I will need to critically assess alternative notions of nature to arrive at a satisfactory definition.

Are humans natural?

The idea that humans are just as much a part of nature as sycamore trees and tigers has certain intuitive appeal. So when developers proclaim that our car parks and roads are no less natural than the woods and meadows, some might find themselves, perhaps reluctantly, conceding the point. What does the claim that humans are natural actually mean?

One reading of this claim is the idea that humans are *identical* with nature. This relationship could be represented in the following diagrammatic way, where N = nature and H = humans:

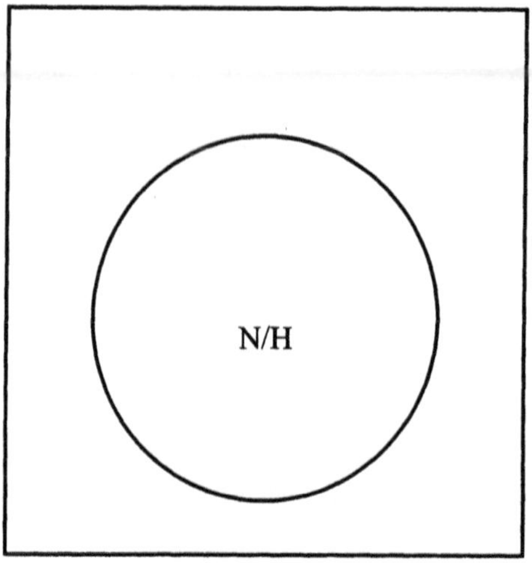

According to Liebniz' law, for A and B to be identical, all of A's properties must be properties of B, and all of B's properties must be properties of A. It is quite obvious that, even if it could be argued that all of humans' properties are properties of nature, it cannot be argued that all of nature's properties are properties of humans. There is, at the most, the possibility of a one-way relationship. In other words, even if it were fair to say that all humans are natural, it does not follow that everything natural is human. We can dismiss, then, any strong association between humans and nature of this kind. Even though the supposition that humans are natural may not be meant as a claim that we are identical with nature, clarification of this point is an important step in revealing its limitation.

Another way of reading the claim that humans are natural is the idea, not that everything natural is human, but only that everything human is natural. In other words, the idea that all humans' properties are properties shared by nature; that humans are nothing *but* nature. This idea can be diagrammatically represented in this way:

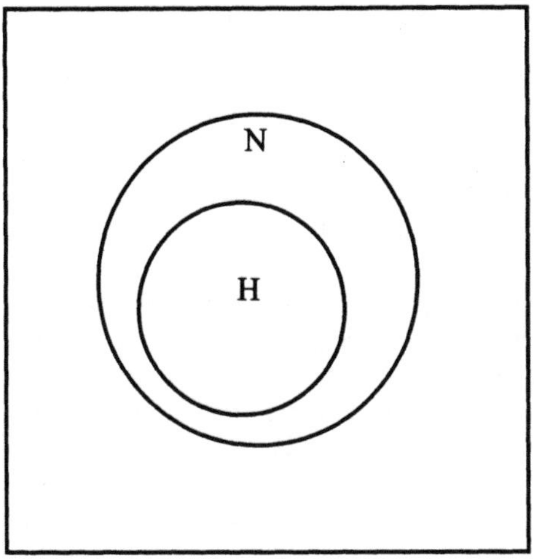

To claim that humans are, as it were, a subset of nature leaves open the question of how nature should be defined. The answer to this depends upon what we contrast nature with. Given this model of the human/nature relationship we seem to have two possible ways of defining nature.

Firstly, 'the natural' might be contrasted with 'the supernatural' in the way that it is defined in J. S. Mill's (1874) broad notion of the natural as a 'name for all facts actual and possible'. Certainly, we are familiar with this distinction. We might wonder, for example, whether crop circles are naturally occurring phenomena or whether alien spacecraft have caused them. However, natural versus supernatural is only one aspect of the term 'natural' and, more pertinently, it is generally not what we mean when we use the term in this sort of context. The natural is more usually contrasted, not with the supernatural, but rather, with the *unnatural*, a contrast that is to be found within the realm of 'all facts actual and possible'. Of course, it could be objected that common usage of a term provides poor evidence for its correct definition. Common usage, however, even if not a determinant of meaning is a good guide. Certainly, it is adequate for our purpose here since we only require that its use is meaningful to us. However, if we are to suppose that humans are nothing but nature, we remain puzzled as to what nature can be. To say that it is 'whatever is not supernatural' seems to include too much, and to say that it is 'whatever is not unnatural' does not enlighten us.

An alternative way of defining nature within the above model would be in terms of evolution: nature is everything that has evolved and is to be contrasted with those entities that have not evolved. There is some evidence that this is a legitimate way of characterising the human/nature relationship. The legacy with which Charles Darwin left us is the insight that humans are animals, having evolved along with all other life forms, through the process of natural selection. As Huxley (quoted in Flew, 1967) puts it, man '"no longer ought to feel separated from the rest of nature, for he is part of it – that part which has become conscious, capable of love and understanding and aspiration ..."' (p. 60). In other words, given an evolutionary reading of ourselves, no characteristics can be identified that set us apart from nature, for all our attributes including 'all moral ideas and ideals have originated in the world'. Darwin's legacy, then, is the understanding that, in Flew's words, 'man is wholly a part of nature' (p. 55).

Though I don't intend to question Darwin's insight it is not clear that it leads us, as Huxley and Flew seem to imply, to the conclusion that we are nothing *but* nature. I will consider this in a moment, but first it is worth asking whether 'the natural' is satisfactorily captured by the notion of 'evolved entities'. Again, it does succeed in correctly characterising an *aspect* of 'the natural'. We would say, for instance, that the evolutionary process is part of the natural history of species. However, we would not say that evolution *per se* is natural, for bicycles, aeroplanes and ideas evolve too. More importantly, though, defining the natural in terms of 'evolved entities' seems too restrictive: what of rocks, waterfalls and tornadoes, are they not part of the natural world too?

If nature is less than successfully defined by the notion of 'evolved entities', there is a problem, too, with the idea that since humans have evolved, we are considered to be nothing *but* nature. I return to Darwin's work in order to demonstrate this.

Darwin's genius was, of course, the postulation of a mechanism – natural selection – for evolution, and he arrived at his theory with the help of the study of artificial selection in domesticated animals (see *The Origin of Species*). As the only species capable of what Darwin calls artificial selection, humans, it might reasonably be argued, must be set apart from the rest of nature in which natural selection operates. Flew quotes Julian Huxley:

> It is in large measure due to Darwin's work on biological evolution that we now possess this new vision of human destiny ... [which destiny] is to be the chief agent for the future of evolution on this planet [for man] finds himself in the unexpected position of business manager for the cosmic process of evolution (p. 52).

In other words, as beings capable of re-directing the evolutionary process, humans can be set apart from the rest of nature in which selection has no direction, no goals. Indeed, our manipulation of the selection process appears to be bound only by our imagination, as the many and varied breeds of domestic animals, and more recently, our advances in genetic engineering, bear testimony to.

All the same, it may be objected that Darwin's identification of artificial selection is not, in itself, evidence of the distinctiveness of humans from the rest of nature. After all, artificial selection is only so-named because of a pre-conviction that humans select in a different way to nature. Perhaps this is a valid objection; perhaps human-directed selection could be considered to be essentially the same kind of process as natural selection – that is, the mechanism by which evolutionary change occurs. Yet the differences between the two processes are substantial, and on balance, there seems to be a good case for singling out artificial selection and setting it apart from natural selection. Briefly, the points of departure from natural selection are:

- Artificial selection is a goal-directed process; natural selection is not.
- Artificial selection is directed by the requirements of a single species rather than resulting from changes that happen to occur in the environment as with natural selection.
- Unlike natural selection, artificial selection has little if anything to do with environmental adaptiveness.
- Artificial selection can effect much more rapid change than natural selection.
- Artificial selection can move outside the realm of natural selection. Natural selection operates upon whole individuals while artificial selection includes, also, direct manipulation of genes.
- Artificial selection enables genetic combinations impossible to produce through natural selection. For example, rabbits can be engineered to glow in the dark through the insertion of starfish genes.

To suppose that artificial selection is a sub-set, as it were, of natural selection, is to be guilty of too comprehensive a reading of natural selection: the differences between the two processes strongly suggest that humans cannot be described as being nothing but nature.

Artificial selection is not, however, the only feature that distinguishes humans from the rest of nature. Humans are set apart, too, in their being, alone amongst all creatures, *moral* beings. This claim may be immediately countered with the assertion that morality is a naturally evolved phenomenon, and perhaps, too, with the suggestion that the rudiments of morality can be seen in, for example, some of the apes and monkeys (see De Waal, 1996). Even so, it does seem fair to say that the formal structure of the human moral system sets it apart, in some way, from its (co-operative behaviour) roots. Whether morality requires the careful direction of natural instincts (as Aristotle thought) or the counteracting of such instincts, morality is something more than the instincts themselves. At the very least, the raw material requires nurturing and honing. Individual goal-directed effort is an essential ingredient of morality, and for morality to operate formally, an understanding of the concept of 'moral system' must have been reached by at least some within that system.

It may be supposed that at least one form of morality – evolutionary ethics – places us back within the realm of nature. Evolutionary ethics holds that we should look to the natural world in order to judge right from wrong action. That is, whatever occurs naturally is right and should be taken as a model for human action. Leaving aside the thought that as a general ethical system, evolutionary ethics seems at the very least, wanting, it might be thought to have particular application to conservation practice. This is because it is generally accepted that right conservation action is determined by, or at least rooted in, 'the natural'. It is quite clear, however, that evolutionary ethics no less than any other ethical system *necessitates* a dichotomy between humans and nature. If we are nothing *but* nature, then whatever action we take must be the right one. It is only when we assume that humans are not *just* nature that it becomes meaningful to distinguish between natural and unnatural, and for the evolutionary ethicist to distinguish between right and wrong action.

Clearly then, to claim that humans are part of nature cannot be to claim that humans are nothing but nature; there are differences between the two. Indeed, some have argued that we are only able to conceptualise and identify nature because we are apart from it. If humans were nothing *but* nature, they say, we would not be able to conceive of that which we identify as nature for it would be invisible to us. Although arguing from the opposite perspective, Evernden (1985) notes: 'The biologist Jakob Von Uexkull has claimed that a medium is invisible to an organism until it is removed from it: water is unknown to a fish until it discovers air' (p. 48). In the same way, we might argue that it is only through our separation from nature that we are in a position to know it. Conceptually, then, there is a difference between humans and nature.

Conceptually, though, there is a difference between alder trees and ash trees, pandas and porcupines, yet all are nevertheless part of nature. As Flew (1967) points out:

> evolution entails change and unless the process of change has turned full circle, this entails difference – which means that ... if A has evolved from B, A cannot be the same as B (p. 58).

The mere fact of natural phenomena is self-evident and does not provide an argument for the claim that humans are to be regarded as separate from nature. However, the differences that I have argued pertain between humans and nature are not merely of this kind. Some features that differentiate humans from nature set us apart because *they are the very antithesis of nature.* They are not just different but *opposite.* The features that distinguish a polar bear from a grizzly bear – or even a python – are not the antithesis of the other animal's features. They are merely different versions of a general physiological model. Conversely, artificial selection and possibly also formalised morality are pulling *away* from nature in just the same way as those artefacts, the existence of which necessitates the replacement of nature. It is for this reason that we can assuredly say that even though humans are natural beings (we have come about through the process of evolution and natural selection) we *cannot conclude that everything we do is natural.* The developers (and others') argument that our car parks and roads are every bit as natural as the rivers, woods and marshes cannot be accepted at face value – it requires justification. Such a claim does not follow from the observation that we are natural beings. This is not only because it requires further argument, but also because when the facts are examined, the evidence suggests otherwise.

I arrive, then, at the conclusion that humans are neither synonymous with, nor a subset of nature. Humans have aspects of them that are outside of or antithetical to nature. We are, however, as Darwin showed, naturally evolved beings and to this extent we and nature are overlapping concepts. This idea could be represented diagrammatically:

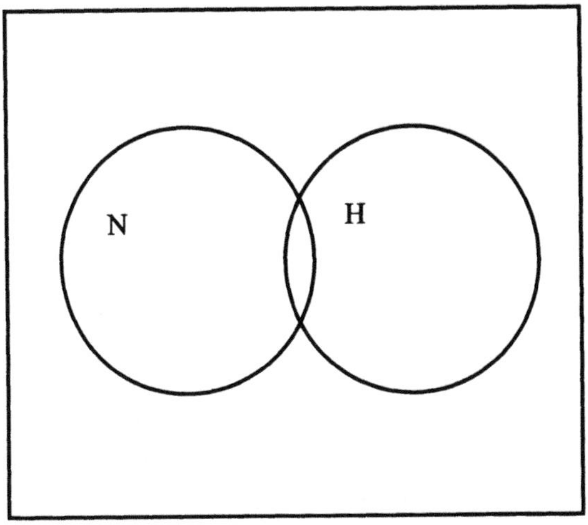

Recalling the features that separate humans from nature let me attempt another definition of nature. J. S. Mill's narrower definition of nature is based on the contrast with human agency. Nature, he says, is 'what takes place without the voluntary and intentional agency of man'. This definition is open to criticism. It is not clear, for example, that only – or indeed all – involuntary or unintentional actions are natural ones. Nevertheless, it is doubtless a closer approximation of what we mean by 'nature' than we have so far encountered. Those aspects of humans – artificial selection and formal morality – that were earlier identified as distinguishing us from nature, have in common the feature of human agency. The combined concepts of artificiality, formal structure and human agency could be gathered under the banner of 'culture', such that culture is the hybrid concept that is set against 'nature'. On one level, 'nature' and 'culture' are polar concepts. But culture must also be understood in terms of a continuum such that points along its trajectory can have more or less in common with nature. Put in another way, if humans are distinguished from nature by way of *contrivance*, there are different degrees and levels of contrivance along the cultural plane. Humans and nature overlap, as I have said, in the area of evolution. But it seems, too, that there is another source of commonality, this being at one limit (the nature end) of the cultural continuum. It could feasibly be the case that, at this point, nature and culture share sufficient common ground as to be distinguishable from one another only in the mundane sense. That is, here, they are not in opposition. Again, a diagrammatic representation is perhaps the best way to capture the essence of these thoughts:

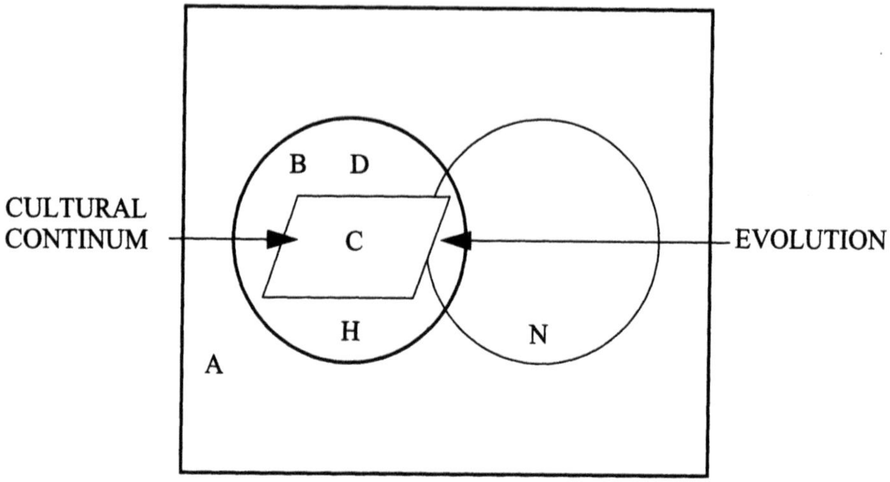

KEY
A: Building such as skyscrapers and roads; genetic engineering.
B: Building such as drystone walls; artificial selection of domesticated animals.
C: Managing urban nature reserves; species reintroduction.
D: Humans living alongside nature, including, for example, simultaneously using and protecting habitats and sustainable living.

So far, my approach has been to demonstrate ways in which humans are distinct from nature, an approach that relies upon our being able to conceptualise and identify nature. As I pointed out, unless we are able to do this, the entire conservation project is meaningless. But perhaps I am too hasty in making this assumption. Indeed, some have questioned whether we *can*, in fact, talk coherently of nature. Evernden (1985) implies that we cannot, by challenging our ability to coherently define the individual. Organisms, he argues, can only be defined *in context*. Firstly, he says, an animal is not just its genes, it is 'an interaction of genetic potential with environment and conspecifics'. He goes on:

> A solitary gorilla in a zoo is not really a gorilla; it is a gorilla-shaped imitation of a social being which can only develop fully in a society of kindred beings. And that society in turn is only itself when it is in its environmental context, and so on (p. 13).

Any individual creature, Evernden claims, is actually a process of relationships.

Secondly, Evernden challenges our everyday claim that we can coherently define the boundaries of individuals, a challenge that gains its strongest support from the existence of those creatures known as symbionts. He writes:

> What are we to make of lichen, composed of two unrelated organisms? The algal component provides photosynthetic ability, the fungal component structure and nutrient intake. The two are mutually dependent. Is the lichen a plant? A co-operative? What about colonial organisms ... (p. 38).

Chloroplasts, within plant cells 'behave independently of the cell' and though 'indistinguishable from the rest of the plant' may have 'originated as separate organisms' in which case, 'what we take to be a single organism may actually be more like the lichen, an instance of mutualism in which the fates of two beings are so intertwined as to make them almost indistinguishable' (p. 39). Even humans are thought to 'share in this kind of ambiguity'. The cellular mitochondria appear to have an independent existence, behaving rather like colonists, such that '[w]e cannot exist without them, and yet they may not strictly be "us"'.

Evernden does not only question whether and how we can draw the line between one creature and another, but also whether there exists a boundary between ourselves and the non-living world; how, that is, we can 'make any sense of the concept of man as a discrete entity' (p. 40). He concludes that:

> What all this suggests is that our assumptions of separateness are unacceptably simplistic, and that we might more closely approximate the facts of existence by regarding ourselves less as objects than as sets of relationships, or as processes in time rather than static forms.

Evernden draws on the ideas of Von Uexkull to depict the organism as 'a very complete interrelation of self and world, so complete that the world could serve as a definition of the self' (p. 81). Evernden sees the 'self' as constituted by something other than that which we generally regard it to be – bounded and identifiable. The notion of territoriality illustrates his point, and here he discusses the territorial behaviour of the stickleback fish:

> It is as if the boundary of what the fish considers himself has expanded to the dimensions of the territory. He regards himself as being the size of the territory, no longer an organism bounded by skin but an organism-plus-environment bounded by an invisible integument ... It is as if there were a kind of field in the territory, with the self present through-out but more concentrated at the core (p. 44).

Evernden is not denying that in identifying nature we set ourselves apart from it. Indeed, he writes that '[t]he environment exists because it was made visible by the act of making it separate. It exists because we have excised it from the context of our lives' (p. 126). His claim is, rather, that in doing so, we have lost our way. In looking at the world in the way we do – reductionist, technological, bounded – we not only fail to see how we ought to interact appropriately with nature, we preclude the possibility of appropriate interaction. As Evernden says of the environmental crisis (a state of affairs conservation practice seeks to remedy):

> It is not a question of our encountering the crisis and resolving it through technology. The crisis is not simply something we can examine and resolve. We are the environmental crisis ... The environmental crisis is inherent in everything we believe and do; it is inherent in the context of our lives (p. 128).

In one sense, Evernden's argument supports the contention that there is a dichotomy between humans and nature. He admits that this is currently the way we view the world. However, if, as he believes, this is the wrong way to understand the world, that 'reality' is, instead, of interconnections and indistinct boundaries, then the idea of a dichotomy would have to be rejected. The question about the nature of reality is a metaphysical one and in order to do it justice I would be required to sidetrack hopelessly. So I must leave to one side this question as well as the question of whether our perception of reality has created the need for conservation. Instead, I must content myself with asking whether it makes sense to us as the kind of being that we are, to talk, as we do, about ourselves as distinct from nature. If it does not, then we must cease to talk of differently evaluating human/nature interactions. We must cease to talk of conservation, and even of nature itself.

One of Evernden's arguments concerns the indistinctness of the boundaries enclosing supposed individuals. From this he concludes that we cannot define creatures as individuals, only as beings-in-a-context. It does seem to be a fact that all living things are interconnected with other living things in some ways. However, it seems too strong to claim that therefore we cannot define individuals. Certainly, there are difficult cases to be found – such as lichen – that genuinely puzzle us. But such difficult cases are to be found around the periphery, as it were, of many definitions (consider, for example, 'art' or 'love'). This does not preclude the possibility of there being examples in which we have complete conviction – for instance, that a human, a song thrush and an elephant are individuals. Indeed, we would never recognise Evernden's lichen example *as* a puzzle case were we not considering it from the individual perspective. The fact that our cellular mitochondria appear to behave as colonists does not readily either exclude them as part of ourselves or rule out the possibility of defining us as individuals. There is nothing in the concept of 'individual' that precludes this possibility. It still makes perfect sense to talk about my experiences as distinct from *your* experiences, and this is particularly obvious with experiences of pain: my pain is not your pain. As Hospers (1970, p. 382) puts it:

> If I feel the pain, it is my pain regardless of where the injury is. In this sense it is logically impossible for me to feel your pain. If I feel it, it is *ipso facto* my pain, and if you feel it, it is your pain. If we both feel pain, it is not the same pain we feel: you feel your pain and I feel mine. If there are two pain-experiences, there are two pains, since the word 'pain' is the label for an experience, and has no existence apart from being experienced.

Indeed, it is the very fact that individuals' experiences are private in a certain sense that creates the 'problem of other minds'. Whether or not our mitochondria have something of an independent existence, my mitochondria contribute to my experiences while your mitochondria contribute to your experiences. That is, what is important in defining an individual is not the way in which their consciousness is constituted but simply the fact of there being a unique, self-contained, non-transferable, identifiable consciousness. Evernden is guilty of overlooking the strongest evidence for individuality: the subjective individual. And even where consciousness cannot be detected in a living thing (plants and some animals) individuality still makes sense to us, granted in a less clear-cut way. Each plant, for example, has its own vascular system on which it depends. For my purposes, here, though, it is only necessary that beings with consciousness be differentiable as individuals.

We do seem to be so constituted as to recognise at least conscious things as distinct entities. And we behave towards them both as distinct entities ourselves (I take responsibility for my actions) and as if they are distinct entities (people – friends, confidantes, lovers, relatives etc. – are not interchangeable or replaceable). It simply is the case that we understand and relate to individuals as beings bounded by their skin. It makes sense to us to talk about individuals and to be treated as individuals ourselves. Nevertheless, this does not mean that Evernden's idea of continuity and inter-relatedness has to be rejected; the concepts are not mutually exclusive. Indeed, the concept of inter-relatedness *relies* on the concept of the individual. Without there being individual entities, there would be nothing to which inter-relatedness could apply. We could concede, too, that whilst an organism is an individual no matter where it happens to be (in its natural habitat, a zoo, a laboratory) it may only be fully itself, fully knowable, in a particular context. A wild animal has the same capacity for suffering – or indeed for any kind of experiencing – whether in captivity or free. In this sense its individuality never changes. However, there is a clear sense in which, for an individual to be fully itself, it requires a particular context that provides particular interrelationships, outlets and opportunities. A caged wild animal is, without doubt, a mere shadow of its free self. But to admit to this is not to be committed to Evernden's line that there is nothing to separate the individual (human or otherwise) either from any other individual or from nature itself.

In contrast to Evernden, Livingston (1994) strongly supports the contention that humans are separate from nature. Indeed, he claims that we are so constituted, as beings addicted to 'know-how' or technique (a fabricated, prosthetic device), to approach and apprehend the world in this way. He says:

> ... the cultural prosthesis is not a mere whim, fancy, or conceit. It is necessary. Our physical and sociobehavioural evolution have brought us to a point at which we cannot do without it. We need it in order to function as social beings ... we will continue to need our supportive cultural prosthesis just as long as we persist in the denial of our animalness, our biological being. Our historic rejection of our own biology, which is commonly expressed as the human/Nature distinction, means that we require an artificial device to replace it (pp. 10–11).

We are, in Livingston's view, the ultimate domesticate, ecologically placeless, alien misfits in the natural world, separate from nature. (It should be noted, here, that Livingston is using 'domesticate' not in its recognised form of a being that has been fashioned through artificial selection, but in a particular way of his own. Humans are domesticates, he believes, because we are fashioned, bounded and thereby excluded from nature, by the cultural prosthesis of technological know-how.) As Livingston sees it, 'were it not for human domestication there would be no "Nature", no "other" no concept of wildness' (p. 36). Livingston takes the extreme view that '[n]owhere may the human presence be seen as fully integrated and "natural", because wherever we may be, or however long we may have been there, we are still domesticates' (p. 57). The human ideology, he explains:

> rests upon a dualistic image of the relationship between human beings and all other natural phenomena. Its stance is human chauvinist; its program is Baconian conquest; its means is Cartesian rationality; its instruments are science and technology.

Livingston talks of our separateness, our ecological placelessness as inevitable, but only because we view the human/nature relationship in a dualistic manner. Should we, instead, see interrelationship rather than boundedness, the self as group, community, even biosphere rather than as individual, there may be some hope of our finding a place for ourselves in nature. Livingston believes:

> There can be little doubt that we shall always specialize in 'how to'. That is the very structure and function of the human brain. [But] ... [t]here must be an alternative way of human being in the world (p. 180).

Livingston's position seems ambiguous. He claims that we are so constituted as to be technology-reliant beings. We will probably always, he suggests, be ecologically placeless domesticates. Yet Livingston is also hopeful that we may yet learn to be part of nature:

> We are different from Nature in that we are domesticated or culturally conditioned, and that we depend upon a fabricated prosthesis to function in an ideology-dependent universe we have ourselves created. But ideologies change. Look at a child gently holding an unfledged young robin that has fallen from its nest. Look at that child's eyes. The sweet bondage of wildness is recoverable (p. 197).

The problem with Livingston's position is that technology (or 'how-to-do') by its nature seems to *require* a view of the world in dualistic terms. It is difficult to reconcile the idea of humans as technology-dependent while also rejecting the dualist perspective. That is, if, as Livingston suggests, we need to first understand our present ideology and then reject it 'together with all of its arrogant, cruel and unnatural appendages', and take on, instead, a new ideology, it is difficult to see how this fits with the belief that we will always be bonded to technology. However, one understanding of Livingston's position in which the ambiguity is removed is the idea that reality is perceived, as it were, as a gestalt image, where there is more than one possible reading of it. Just as we are able to switch from recognising one image to another in the gestalt picture, we can move from a dualistic understanding of the world – that is, where there is a human/nature dichotomy – to one in which humans are understood as continuous with nature. Certainly, psychologically, we can move from a perception of self as individual to self as group or community. We can appreciate Livingston's sentiments in his talk of the group-self of a flock of grackles:

> In highly gregarious species such as flocking birds, the individual is lost not only to our sight but also to our awareness ... the idea of a single grackle, though conceivable, has no reality at all. The harsh grating cacophony of their voices, the noisy wave of flappeting wings in the snowy branches, bespeak the quality of grackleness (p. 104).

We might even understand what Livingston means when he says that '[t]he individual grackle has for me no meaning – at least at this moment – whatever', but I for one would want to point out that presumably this is not the case for the individual grackle! The important point is that this way of perceiving the birds is just one way, and transient at that. It can never *replace* the individual perspective. We would shift our perception to one of self-as-individual should we wish to study their behaviour or should one bird fall injured to the ground. Livingston is surely right to highlight the way in which a perception of the inter-relatedness or bondedness of nature heightens our pleasure and feelings of participation. However, he neglects to acknowledge the powerful effect that nature can have upon us when it is revealed to us in intimate and unexpected ways. The experience of handling a grounded wild bird – noting how very different it looks in the hand from its appearance in the air, feeling its warmth, watching its rapid breathing and palpable fear, its sheer aliveness – is a vivid reminder that in nature, self-as-individual is both a reality as well as an inspiration.

It is not only, then, that a dualistic perspective is a reality; it is more than 'just one way' of perceiving ourselves and nature. In contrast with Livingston's view that it perpetuates the environmental crisis, I suggest that a dualistic perspective has a valuable part to play in creating a *better* relationship between nature and ourselves. The individualisation of ourselves and others is an important aspect of the kind of being we are. In that this is so, when nature is understood and appreciated in this way, it can bring us closer to it, which in turn offers promise as a route through which the so-called environmental crisis may be solved.

I am not arguing that a holistic understanding of the world makes no sense at all to us: certainly, it is meaningful. I am arguing, however, that it neither eclipses nor displaces an individualistic understanding of the world. It has its place, then, as a perspective that adds depth and richness to our view of the world, but cannot and should not replace individualism. To the objection that this conclusion cannot be taken seriously because it is reached by working within the individual paradigm, the only response possible is the observation that we have to approach the question from within *some* paradigm. To work within the holistic paradigm would be equally open to the charge of bias. The best that can be done – and which I have aimed at here – is to assess the strengths of argument from either side from within a chosen paradigm (indeed, the arguments could not otherwise be compared) and thereby reach a conclusion. The individual paradigm is, I believe, the most appropriate choice for the kind of being that we are.

I can conclude, then, that although there is a real sense in which humans are part of nature, it also makes sense to talk about us as separate from nature, and therefore to talk in terms of an interaction. However, since we can interact with nature in a myriad of ways, the question, now (given the assumption that we wish to safeguard nature) is what kind of interaction is most desirable.

PART 2
Humans and nature: is intervention desirable?

The conservationist's aim is to safeguard nature, however there is much disagreement as to how this is best achieved. Disagreement is to be found at various levels, the most fundamental of which is whether or not we ought to intervene, and this is (chiefly) because human activity per se is regarded by many as the antithesis of nature. There is seen to be a tension between the ideas of nature and human impact. Human impact is believed to damage, degrade or in some way compromise nature. I have already suggested – in the idea of a cultural continuum – that human impact on nature need not lead to such an outcome. I suggested that there are varying degrees of artificiality, or cultural contrivance. But I propose now to consider more fully the question of whether – and to what degree – human intervention is compatible with the conservation of nature. I will do so by examining three possible positions that the conservationist might take.

The 'purist' non- interventionist

Given that the need for conservation has arisen because of inappropriate human intervention in nature, it may seem quite plausible that the best way to safeguard nature would be a 'hands-off' non-interventionist approach. This is the thinking behind the designation of wilderness areas in North America, in which human presence is tolerated provided we are merely benign and transient visitors. This 'purist' position has undoubted appeal and on first consideration may well seem to be the way forward for conservation. However, after more careful scrutiny, it becomes clear that a non-interventionist position is problematic. Here, I will consider three serious objections.

Firstly, it could feasibly be argued that the idea of safeguarding nature through non-intervention is paradoxical. In order to ensure the safeguarding of an area of nature, some kind of measures have to be implemented (even if simply the action of setting land aside) and this constitutes intervention of a kind. As Vogel (1996) points out:

> ... even 'true' wilderness remains so only by virtue of a complex set of social mediations: the acts required here are Acts of Congress, by which we socially decide to remove areas from the sphere of transformation, thus revealing the extent to which for us wilderness is now founded upon and carved out of the world of the social and not vice-versa (p. 37).

There is, then, a logical difficulty with the non-interventionist position.

There is a metaphysical point to be made too. If, as Vogel believes, 'reality' (including the reality of nature) is to be understood as a social construction, then it could be argued that we have 'intervened' in nature at the most fundamental level possible. Nature, as Vogel argues, 'never appears as it "is" but rather always as already the nature of a particular social order, in a particular context, subject to a particular set of mediations' (p. 37). But even considered from a practical perspective, the notion of non-intervention is problematic. How are humans, located as we are in and around nature, to *avoid* affecting it? Granted, the idea of safeguarding nature must take on board the importance of reducing negative human impacts (and in this, all conservation positions will accord with non-intervention). But given the nature of nature, in which the smallest occurrence can have far-reaching effects, it is clear that even the most benign of our activities cannot fail to have an impact upon nature, whether we are excluded from it or not.

The above objections confront the non-interventionist position with genuine difficulties. However, they may not cut much ice with the practical conservationist who is concerned not with logical or metaphysical points, but with the question of how to safeguard nature. The bottom line for the conservationist is whether or not a non-interventionist strategy is effective. Here lies the second difficulty. A strictly non-interventionist position requires that humans are excluded from those tracts of land to be safeguarded. This is a strategy that lends itself to some countries (the USA and Canada, for instance) but not to others (Britain and France, say). Certainly in Britain, there are few areas that are entirely uninhabited by humans. The John Muir Trust, a prominent British conservation organisation, aims 'to safeguard wild land'. The Trust's namesake, John Muir, wanted to 'do something for wildness and make the mountains glad' and he considered that the way to achieve this was to ensure the integration of humans and wild land. The Trust's information leaflet explains that:

> People have lived here for thousands of years, and left their mark everywhere. Little of our landscape is unchanged. The John Muir Trust is committed to working closely with people who live on Trust land, sustaining local communities while maintaining the special qualities of wild places.

In Britain, at least, non-intervention is not a feasible option for the conservationist. And if the John Muir Trust is a good guide, neither is it a necessary or even desirable option. Such a strategy cannot, then, be regarded as a general conservation strategy: its potential for application is too exclusive.

Even in places such as North America its application is questionable. As a strategy that is bound to exclude humans it has limited use. It may be argued that the implementation, here, of protected inner core areas with outer buffer zones in which limited and benign human activity is allowed, succeeds in combining non-intervention with a realistic appreciation of human predomination. However, this is not a pure non-interventionist approach: if it succeeds, it does so because it is a combination of non-intervention with intervention. The objection that buffer zones are merely an interventionist means to a non-interventionist end is not wholly convincing. It still leads us to question whether nature conservation can be achieved *through* non-intervention.

A third and fairly damning objection to the pure non-interventionist position is that it is self-defeating. There are times when the aims of the non-interventionist would be much better achieved by intervention. One such example would be in the case of an accidental oil spill. The removal of the oil, the cleaning of oiled wildlife and so on is likely to be the best way of achieving pre-spill conditions. Yet such intervention could not be sanctioned by the strict non-interventionist. This position does not allow for human action directed towards ameliorating the results of accidental or incidental, undesirable intervention.

The pure non-interventionist position is too problematic to be taken seriously. We must look elsewhere for a sound conservation position.

The 'modified' non-interventionist

One possible way of avoiding some of the problems inherent in the strict non-interventionist position is the adoption of a modified non-interventionist position in which the cut-off point that legitimates human intervention is one of 'obvious harm' such as oil pollution. One way of characterising this position would be to say that intervention is permissible if it seeks to remove a previous intervention but not if it seeks to intervene in any other way. It is acceptable to 'undo' but not to 'do'. This may seem a very reasonable stance and a good compromise: there is nothing wrong with human intervention provided it seeks to redress the harm already done. Compelling as it may be, though, the modified non-interventionist position has serious defects, three of which I will consider.

To begin with, there is arguably a certain tension within the concept of the modified non-interventionist position, in that the ideal (nature ought to be untouched) does not sit comfortably with the process (intervention) by which it is attained. Perhaps, though, this objection lacks any real substance since it relies, for its impact, upon terminology. That is, the objection could be countered simply by re-naming the position. At the very least, it could be argued that the notion of 'undoing' – which would seem to be an act of removal – conjures an image of humans retreating from nature that is compatible with non-intervention. But this image is tarnished by the recognition that sometimes the most effective (or perhaps the only) way of 'undoing' (or neutralising) an earlier action is not removal of but *addition* to that which is unwanted. Two examples will suffice: an effective method of dealing with oil pollution is the addition of oil-ingesting microorganisms, and 'pest' species are sometimes best dealt with by the introduction of a predator. When the concept of 'undoing' is as broad as to include the addition to nature of the human-dictated, the novel or the engineered, the idea that tension arises in a position that advocates non-intervention through a process of intervention gains credence.

Whilst the notion of 'undoing' (or neutralising by removal of) that which is considered harmful to nature is appealing, it may not make sense in ecological terms. In nature, once change has taken place, it is simply not possible to return to the former state of affairs because things will have moved on. That is, the very fact that change has taken place precludes the possibility of experiencing the pre-change circumstances again. When change occurs, a new set of processes, inter-relationships, opportunities, are set in motion such that any present or future trajectory will necessarily differ from any past trajectory, no matter how much effort is made to redress and restore. This is not to say – for example following an oil spill – that it would be pointless to attempt to clean up the oil. It is only to point out that the idea of returning to past conditions is conceptually unviable and therefore to problematise a position that fails to recognise this.

The position of the modified non-interventionist cannot rest solely on the concept of 'undoing'. Were we to attempt to undo *all* of our actions that impacted upon nature we would be unable to function, for however we live we have an impact upon nature. It would be an impossible task to reverse all human impacts – not only practically, but also conceptually – because 'undoing' itself imposes an impact. The position requires that we can differentiate between desirable and undesirable impacts and that we seek to reverse the latter. The problem, here, is with the definition of 'undesirable impacts', or, as I earlier suggested, the notion of 'obvious harm'. Conservationists seem to suppose that environmental degradation can be straightforwardly recognised, but this is far from the case. Let me outline some of the difficulties with this concept.

The notion of 'harm' or 'degradation' must be differentiated from the notion of 'change'. All too frequently the two are conflated. It is not clear that when woodland becomes open moorland, or savannah becomes desert, that *degradation* has occurred any more than when moorland becomes woodland, environmental enhancement has necessarily taken place. Vegetation types have changed continuously throughout history, some as a result of human impact, others due to naturally occurring phenomena, some welcomed others not. There is no direct correlation between change and degradation.

The question of whether a change is harmful or not requires us to ask 'from whose point of view?' It may be widely accepted that an oil-spill is a paradigm case of environmental degradation. However, at least one organism – an oil-ingesting microbe – thrives in such an environment. Similarly, an environment depleted of oxygen may well be considered to be a degraded one, but this would not be so considered from the perspective of anaerobic organisms. As a third example, global warming that is occurring as a result of our pollution is regarded as environmentally harmful. However, in its warming effect on the climate, we can expect to see 'increasing numbers of insects, including a bigger variety of species in the north' (Neale & Dean, 1998). The point is that in assessing whether change denotes degradation, conservationists may be blinkered or biased. It may turn out that the notion of degradation is based upon some kind of numerical calculation: if the change disadvantages more organisms than it benefits then it is considered to be a harmful one. Alternatively, degradation may be assessed in terms of advantage or disadvantage to humans or some group of organisms we deem to be important. There may, then, be no impact that is, *by its nature*, a harmful one. The categorisation of environmental degradation is, at least to some degree, a relational matter.

Even if there is wide agreement that certain impacts represent destruction, or degradation, it may be the case that some such impacts are considered to be environmentally valuable. Tickell (1995) points out, for instance, that woodland benefits from the damage caused by large animals crashing through the vegetation and churning up its floor. Fires are sometimes valuable regenerators and storm-damaged trees provide important habitats for numerous species. So even under the category of 'damage', the question of whether or not the damage is undesirable is relative or context-dependent.

One definition of environmental harm, popular amongst conservationists, is 'loss of diversity'. On its own, however, the notion of biodiversity cannot be considered a value. If this were so, it would always be the case that the more 'types' (sub-species, species, habitats etc.) the better. It would also always be the case that low-diversity habitats are less valued than high-diversity habitats (notwithstanding that low-diversity habitats would still be valuable, presumably, in their contribution to habitat diversity). The conservationist, however, considers neither of these to be so. But if 'loss of diversity' as such cannot be considered to represent environmental harm, then loss of the *opportunity* for diversity may do. Recalling Chapter One, the current anthropogenic extinction event has a profoundly different effect upon the natural world than did previous extinction events. It appears to be cutting off the opportunity for speciation – or the perpetuation of diversity. It seems reasonable to suppose that to cut off the *source* of our natural world is to harm it. However, if this represents the cut-off point that legitimates human intervention, it is not clear that it prohibits very much. All sorts of intervention might be justified in order to maximise the opportunity for speciation – such as culling, species reintroduction, genetic engineering, and many other management strategies. So, the one definition of environmental harm that is meaningful cannot hope to be a guide for a modified non-interventionist position. It simply legitimates too wide a range of interventions. To call a position based on this criterion non-interventionist would be nonsensical. Similarly, even if some agreement could be reached about which specific impacts of ours constitute 'environmental harm' (e.g. if we came to a consensus about what *pollution* is) it is likely that these harms are so all-pervading as to legitimate a degree of (pollution-reversing) intervention that is incompatible with any notion of non-intervention.

It seems, then, that though non-intervention might at first be supposed to be the best way to serve conservation's aims, it is not a viable position, either in its pure form or when modified to allow intervention for the redress of environmental harm. The pure non-interventionist position has logical, metaphysical, and practical problems and is ultimately self-defeating. The modified non-interventionist position fares no better, since there is a conceptual tension both in terms of the internal structure as well as the way it operates in the world. Indeed it became clear that the position's fundamental concept – that of 'undoing' – is unworkable. The alternative, interventionist position may seem to be no better in that it appears to be at odds with conservation's goal of safeguarding nature. I am going to consider whether this is, in fact, so, or whether it might, after all, offer the best conservation approach.

The 'thinking interventionist'

The notion of an interventionist position is impossibly wide; it legitimates human impact of any kind. Clearly, a position such as this could not possibly serve to fulfil the aims of conservation. Indeed, as I noted earlier, conservation is a response to inappropriate human impacts upon nature. There is, however, a modified version of this position that, I propose, offers just the right approach for determining how we should live with nature to best conserve it. This is what I am calling the 'thinking interventionist' position.

The 'thinking interventionist' position, though clearly pro-intervention, is characterised by the *thoughtful* interventionist, a concept that captures not only the rationality of the position but also the notion of care. In other words, the attitude of the conservationist is of primary importance. The broad thrust of the position is one of measured intervention, that is, where intervention is directed by careful consideration and good reasons. A distinctive feature of this position is its reference to the theme of 'the past' in which raw nature (minimum human impact) becomes the model or touchstone on which decisions regarding intervention are based. Interventions, though of central importance in themselves, could only be sanctioned, then, if they honoured, as it were, the theme of raw nature. This is not to say that intervention should seek to mimic, or return to the past (notwithstanding, as I have said, that this is an unattainable aim), but rather that raw nature should always be referred to as a model when intervention is being considered. The remit of the conservationist would be to ensure that whatever changes took place, the 'new' nature maintained substantial links with the 'old' (raw) nature. In short, there would be a requirement for a strong sense of *continuity*.

It is clear that such a position will positively favour actions of a redressing kind, such as the clean up after an oil spill. And the need for redressing action (as well as the definition of environmental harm) will be determined by the degree to which an impact deviates from the raw nature model. However, the 'thinking interventionist' position does not only concern itself with redressing action. Other impacts that seek to effect new changes will also be sanctioned, provided they are in keeping with the past. This is similar to the way that new housing is permitted in the Lake District National Park only when it accords with the style and structure of existing dwellings. In that new impacts are permissible, the 'thinking interventionist' position demonstrates a healthy respect for what has been termed 'future nature'.[1] This is the antidote, as it were, to the stifling non-interventionist conservation position in which nature is treated as an ark, to be preserved, as if in ice, in some designated state. To make room for future nature is to allow opportunity for new processes, inter-relationships, and so on, that may not have occurred previously. It is to allow a new, as yet unknown trajectory to unfold. This strikes a chord with Sylvan's idea of environmental restitution (alternatively described as rehabilitation or restoration) about which he says: 'Restitution need not ... be backward-focused or regressive, any more than deindustrialisation. Both can be progressive with a strong appreciation of the past and of past mega-mistakes' (Footnote 18, p. 19).

It might be said, then, that the value of innovation is determined by its degree of continuity with raw nature. To be more specific, the 'thinking interventionist' position would not, for example, sanction the replacing of a natural or semi-natural habitat with something other than a habitat (for example a car park) for there would be too little continuity with raw nature (they would have little, if anything, in common). Such an innovation would have no conservation value. But the 'thinking interventionist' position might sanction the change, say, from habitat A to habitat B by way of innovation, where the innovation allowed for the unfolding of a future nature that was in accord with raw nature. Some concrete examples would be: the reintroduction of species by way of innovatory captive breeding programs (the return of species present prior to human changes would be a powerful link with raw nature); the creation of reed-beds, where none had been in the past, to help the declining bittern (there has been a long association between bittern and reed beds in Britain); and the planting of coniferous woodland to allow the red squirrel to prosper in the face of competition from the grey squirrel, even though the red squirrel/ coniferous woodland combination has no historical precedent in those areas (red

---

1 The idea of future nature is extensively covered by Adams (1996).

squirrels were well-established inhabitants of British woodland prior to both woodland destruction/over-use, and grey squirrel introduction).

I earlier mentioned the objection that in ecological terms, a return to the past is not possible and so in this sense, it could be argued that the unfolding of future nature is *inevitable*. There are, however, different ways of emulating the past, some of which allow more room for future nature than others. Seeking to 'undo' rather than to recreate is one way of stifling future nature. Recreation itself allows for different degrees of future nature to emerge. Factors that influence this include how closely the past is emulated as well as the predictability and degree of control we have over the outcome. These factors are, in turn, linked to the attitude taken towards human intervention. Since intervention is at the very heart of the 'thinking interventionist' position, novelty and innovation are positively welcomed, thereby making for a rich future nature. The 'thinking interventionist' position is, then, one that is buffeted on two sides. On one side, a model of raw nature acts as a touchstone and guards against excessive change or the idea that for an interventionist position, 'anything goes'. On the other side, the idea of intervention is embraced such that innovation and novelty are allowed expression.

The remit of the 'thinking interventionist' is to honour nature's history. But if a sense of continuity is to be maintained, it will be necessary to honour too, the process of human history, because it is through our own history that we reach back to raw nature. It is also through the process of our history that we have come to understand nature in the way that we now do. The continuity between nature today and nature in the past is interwoven with a temporal continuity in which human history features. As Holland and O'Neill (1996) write, the problem we are faced with in making conservation decisions is most usefully construed as 'the problem of how best to continue the narrative' (p. 3). The question we should ask, when wondering how to deal with earlier human impacts is 'what would make the most appropriate trajectory from what has gone before?' Rather than the unrealistic and unworkable notion of non-intervention that seeks to safeguard nature by removing us from it, these sentiments recognise that nature can only be truly safeguarded when we learn to live well *with* it. So those past human impacts that were not in accord with raw nature may be transformed by way of an appropriate trajectory. They may even offer an outlet for the kind of innovation that generates a fully mature future nature. Consider, for instance, old mine shafts left in situ as breeding sites for common frogs (proposed for the Middleton Wood project, Heysham in Lancashire), introduced trees that support a whole array of micro-fauna, and gaps on the underside of bridges deliberately left for use by hibernating bats.

I could summarise the 'thinking interventionist' position as one that takes on board both the idea of minimum human impact (raw nature) and the importance of opportunity for change (future nature). These two ideas come together in the notion of an *appropriate trajectory* for nature. I have suggested, above, some ideas of what might be considered to be appropriate trajectories. Yet the notion of 'appropriateness' may still be considered to be vulnerable to attack. Wittgenstein, in his *Philosophical Investigations*, for example, points out that for any series of numbers, there is an indefinite number of ways of continuing the series. One way cannot be considered to be more appropriate than any other. Similarly, it could be argued that there is an indefinite number of ways of continuing the nature narrative and that it is not obvious why one trajectory should be more appropriate than any other.

The argument I have been advancing here is not that there is one and only one appropriate trajectory for nature. Rather, there are boundaries within which appropriateness lies, and a number of possible scenarios might be deemed indistinguishable in this respect. Nevertheless, the question now moves to one concerning boundaries or limits of appropriateness. How would we know when an action is appropriate (in accord with conservation) or not? What degree of continuity with raw nature is sufficient? Clearly, it is not possible to give a precise prescription for appropriate action, because every case will exhibit different characteristics. Whether or not an action is appropriate is a matter of *judgement*.

Some cases will be easy to judge. For example, the proposal to build a railway up the Cairngorm Mountains in north-west Scotland would be straightforwardly rejected as a project with no conservation value. A railway has no connection with raw nature, and in this location it has no historical significance: the innovation is insufficiently grounded in nature itself. Equally easy to judge would be the reintroduction of the beaver to Britain. This would be considered an appropriate conservation action since the innovation maintains strong links with raw nature as well as with human history (we once interacted with the beaver). At the same time it allows room for future nature to emerge: we do not know whether the reintroduced beaver will thrive or how far it might re-colonise the country, we would simply be creating the opportunity for a new direction in nature. The trajectory of the reintroduced beaver would certainly be different from that of the beaver of the past.

Other cases will be more difficult to judge. For instance, it might be that certain species reintroduction projects are in danger of being too innovative, thereby sharing too little common ground with raw nature. The attempted reintroduction of the whooping crane in North America involved, amongst other things: enclosure of wild cranes with the resultant loss of some wild behaviours including the ability to copulate; the clipping of wing tendons to prevent the birds from flying; various techniques to encourage copulation and breeding; artificial insemination; artificial incubation of eggs; and the largely unsuccessful reintroduction of selected youngsters to the wild (De Blieu, 1993; pp. 105–119). The degree of intervention and the nature of innovation do not sit entirely comfortably with the model of raw nature. Nevertheless, there is clearly a good degree of continuity in that it seeks to re-establish an element of nature.

Opinion might, then, be divided as regards the appropriateness of some actions. There are no formulae for making judgements that are, to varying degrees, affected by culture and fashion. Whilst there are clearly difficulties with arriving at conclusions, nevertheless, there is no alternative to using judgement if actions are to be assessed; we simply cannot avoid it. Yet despite these difficulties, judgement is not a purely subjective exercise. Judgement may be well, or poorly made (see, for example, Holland, 1995). Particularly when exercised in the light of the 'thinking interventionist' model, there seems good reason to suppose that judgement can produce both appropriate action and agreement in the majority of cases.

One final worry with the 'thinking interventionist' position is with the notion of future nature. If future nature is the opportunity for change, for the unfolding of new trajectories, then does it not necessitate the *absence* of human intervention?

Certainly, standing back and allowing nature free reign is one form of future nature. For example, we might stand back and allow erosion to occur, or cease culling deer – and wait to see the outcome. Indeed, sometimes at least, the most thoughtful action the 'thinking interventionist' could take is to do just this – to perform *no* action at all. However, if this were the only permissible kind of action it would clearly amount to a non-interventionist position, which, as I have already shown, is unsustainable. When, instead, non-action is chosen from within the 'thinking interventionist' position, it is thoughtful, it is considered, and it is one possible alternative amongst many. It is not universally, and unthinkingly applied. And even if non-action is not often the best conservation decision, it may be that it is always worth *considering*. David Russell (1998) of the National Trust suggests that a question worth asking on every occasion is 'what would happen if we did nothing?'. It is in answering this question that we better understand what action we need to take. The point is that the 'thinking interventionist' has the *option* of non-intervention – of tapping into this source of future nature – if appropriate to do so. However, when intervention is required, as it very often is, future nature is not necessarily denied expression. Recalling that future nature is the antidote to an ark-like preservation of nature, that it is the opportunity for new possibilities, it is clear that human innovation can provide this. Even though appropriate action is determined, in part, by past or 'raw' nature, there is still sufficient room for the development of new trajectories. A re-invention of the past is significantly different from a desire to return to the past as is exemplified by the 'modified non-interventionist' notion of 'undoing'. It is not to be supposed that conservation decision-making is an easy process. However, by engaging both the rational and caring aspects of thoughtfulness, the 'thinking interventionist' is well placed to make good decisions.

Conclusion

I argue in Part One that it is meaningful to think of humans as both connected to nature as well as separate from it. In an evolutionary sense, we belong. In the sense that we have identified nature and a need for conservation – and in that we exhibit characteristics that are antithetical to those of nature – we must, necessarily, stand apart. Even to have identified our interconnectedness assumes a separateness of some kind; we are not a seamless whole. Individualism is a notion that we are so constituted to employ in our understanding of the world; we see ourselves as individuals and recognise individual others. Individualism supposes separateness both from other individuals as well as from the environment. While we can recognise and appreciate that individuals cannot fully be themselves when removed from their environment, regardless of where found, they remain bounded by their skin and limited to their own set of experiences. In that humans are separate from nature, we can legitimately talk of human interaction with, and impact upon, nature.

In Part Two, my task was to discover how we should interact with nature to best protect it. I found non-intervention to be an unviable position, either in its pure form or modified to legitimate intervention of a redressing kind. The position I suggest as that which best embodies the conservation ideal, is that of the 'thinking interventionist', a combination of conservatism and innovation. It succeeds where other positions fail, and its strengths can be summed up in the following three points.

Firstly, conservation decision-making has often been shown to be biased and arbitrary. This is because there are no real guidelines as to what will best serve its aims. Science is claimed to be a good arbiter in conservation matters. This may be because it is thought that an appeal to science is fully objective. However, according to numerous critics of science (such as Chalmers, 1978) we cannot rely on this. One criticism is that it is impossible to detach an objective science from the practising of science that is context or culture dependent. From the questions that scientists choose to ask to the methods of enquiry, science, it is often argued, is imbued with societal norms. It would seem that science must fail us too in conservation decision-making. As I have argued here, at the heart of conservation lies not merely rational thinking but passion, respect and deep feeling – emotions that science cannot account for. The failure of the scientific approach in addressing conservation matters is emphasised by Shrader-Frechette and McCoy (1994). They advocate that ethical rationality and not scientific rationality should be employed in setting conservation priorities. The latter, they say, makes no room for social, aesthetic and cultural considerations, nor takes account of rights, duties, public preferences and the common good, all of which are values relevant to the human enterprise of conservation. The great strength of the 'thinking interventionist' position is in providing an internal mechanism, a structure for arriving at conservation decisions that acknowledges both the importance of the human element of the conservation quest (the importance of judgement) as well as the importance of natural guidelines (the models of 'raw' nature and future nature). Indeed, it has been convincingly argued that when personal judgement rests on good reasons, it is quite properly non-arbitrary and objective (Holland, 1995). The 'thinking interventionist' position offers scope for making decisions of this kind.

Secondly, the 'thinking interventionist' position allows the emergence of a rich future nature otherwise denied by those positions that seek a static preservation of nature, or a return to, rather than reinvention of, the past. Recall, from Chapter One, that the safeguarding of nature requires, above all else, the opportunity for speciation. It is clear, then, that the unfolding of future nature – of which the formation of new species may be the paradigm case – is of fundamental importance in nature conservation. In making room for future nature, the 'thinking interventionist' position succeeds in honouring the past without this leading to a stultifying, ark-like requirement for nature.

And lastly but vitally, the 'thinking interventionist' position deals with the safeguarding of nature in a realistic way. Indeed, much of what is advocated already accords with conservation practice. The 'thinking interventionist' position does not offer a too radical or unattainable solution to the so-called environmental crisis. Given the fact that humans and nature *must* interact, non-intervention, as a conservation strategy, is only window-dressing and does not get to grips, in any tangible way, with how to safeguard nature. The 'thinking interventionist' position acknowledges that the only way to truly save nature is to find a way for humans to live well *with* it.

Chapter 5

How is nature to be safeguarded? Conservationist, custodian of wildness

Introduction

In Chapter Four I explored the connections between humans and nature and discovered that it makes perfect sense to talk about a relationship between nature and ourselves. Whilst there is clearly a sense in which humans are part of nature, nature must also be understood as standing apart from us, as something with which we interact, or have an impact upon. The idea that conservationists both could – and needed to – conserve nature was shown to be meaningful. However, not only do we interact with nature, we can choose *how* to interact, and in this way help to determine what impact we will have. Since conservationists are at the forefront of environmental policy-making, their position carries with it great responsibility. There is widespread agreement that the role of the conservationist is to safeguard nature as we know it (a nature that, though clearly altered by humans, still carries the hallmark, as it were, of 'raw' nature). However, we need to translate this into a goal towards which the conservationist can aim. In other words, we need to know how the conservationist can best safeguard nature. And since 'the continued opportunity for (natural) speciation' has already been identified as a key factor in the perpetuation of the nature we value (see Chapter One), the conservationist's overarching goal must embody this. With this point in mind, I propose that what the conservationist ought to be working towards protecting and perpetuating is wildness. Indeed, I would suggest that wildness is central to the *meaning* of nature conservation. There are, then, two tasks ahead. The first is to clarify what is meant by 'wildness', and the second is to justify my claim that conservation is best carried out by perpetuating wildness.

The wild experience

In order to articulate what wildness is, it will be helpful first to place it in relation to the more familiar notion of wilderness. Some suppose wildness to be equivalent to the North American notion of wilderness, and indeed, there are authors who use the terms interchangeably (see Katz, 1992). Wilderness is characteristically understood as those places in nature that are devoid of human influence, or, as Des Jardins (1997) puts it, 'an area unspoiled and undisturbed by human activity'. He notes that '[t]he Wilderness Act of 1964 defines wilderness as those areas "where the earth and its community of life are untrammelled by man, where man himself is a visitor who does not remain"' (p. 149). Although it would clearly be true that these places are also wild, it is not true that *only* such places are wild. As articulated in Chapter Four, it may not be possible to identify any place that has escaped human influence, a possibility that problematises the above notion of wilderness. As Des Jardins remarks:

> Humans inhabit much of the globe and human activity affects the entire earth. The climatic and atmospheric effects of pollution is just one example of human activity that reaches every area of Earth. Most wilderness areas are human constructs, in that constant human activity is necessary to preserve and manage them.

The idea of wilderness as a place entirely untrammelled by humans is problematic. Our idea, though, of what passes as a *wild* place tends to be less exacting. It does not, for example, ordinarily exclude the possibility of human influence. For instance, even though humans from Neolithic times onwards have altered all land in Europe in some way, certain areas are nevertheless regarded as wild. It is often conveniently forgotten, too, that some, at least, of North America's wilderness areas were once altered by the native Indians (Holland and O'Neill, 1996). If it emerges that none of the New World's wilderness areas are, in fact, untouched, the appellation 'wild' does not seem inappropriate. Similarly, the Australian bush – extensively burned by the Aboriginal peoples – would still be considered wild by most. And few (if any) would dispute that the Amazonian forest is wild. Though not drastically altered, it has been lived in and influenced by native peoples for a very long time. Places we consider to be wild, then, are very often – if not always – places with a history of human influence. That human intervention can and does deplete wildness is not denied; all that I contend is that wildness is usually considered compatible with certain human impacts. Wildness need not, and often is not, seen to be compromised by the mere fact of human impact. Our ordinary understanding of wildness is one that accepts the legitimacy of human involvement.

It does seem, then, that we are capable of recognising wildness, that we can, at least sometimes, make a distinction between those human impacts that compromise and those that are compatible with wildness. Nevertheless, the notion of wildness is not amenable to being precisely defined, because it is an *experiential* phenomenon. We *feel* rather than logically or empirically determine wildness. When we experience wildness, we are experiencing not just wild entities, but also certain *effects* that help constitute wildness. Grumbine (1994), for example, describes wildness as 'the force or "rules" behind places' (p. 231). A wild place could be said to be the end product, or concrete manifestation, of Grumbine's 'regenerative force' of wildness. Yet even though wildness is, in essence, experiential or phenomenological, we can, to some degree at least, pin it down. There seem to be several aspects that contribute to the full experience, some of which are more concrete than others. Let us turn, now, to consider each of these.

The predominance of nature Perhaps the most obvious and fundamental requirement of wildness is the preponderance of nature over human artefact.[1] That is to say, part of the meaning of wildness is its correlation with nature. We have seen that highly altered landscape can still be considered to be wild and therefore that wildness is compatible with human artefact *per se*. But wildness is not compatible with any, or all, human artefact. There is a point at which artefact eclipses nature and where wildness, as it were, runs out. It might be that human artefact is too extensive; that is, it covers a very large area. Or it might be that human artefact is less extensive but of a particularly invasive kind. Sizewell nuclear power stations in Suffolk may be one such example. Such artefacts are, as it were, antithetical to nature (see Chapter Four). In order for nature to predominate, it must do so, not only in a quantitative sense, but also in a qualitative sense. But there is no formula for determining when either nature or human artefact predominates; all we can say is that in order for wildness to be experienced, there must be a feeling that nature predominates over human artefact.

Necessity for life In arguing that wildness requires the predominance of nature, I raise the question of what is meant by 'nature'. It might be supposed, for example, that the mere absence of human artefact is a sufficient criterion for the identification of nature. But the notion of nature (and by extension, the notion of wildness) carries with it something more than this: it suggests, also, the presence of life. In Chapter Four I said that when we talk of nature as 'natural' the usual (implicit) contrast is with that which is 'unnatural'. The antithesis of the unnatural, the artefactual, the manufactured, is the living organism. Whilst it may be appropriate to describe the moon, say, as wild,[2] it would not seem appropriate to describe it as 'wild nature'. It has, to our knowledge, no life upon it. Nature, then – and by extension wild nature – is lifebound.

Systems of relationships In order for wildness to be experienced, the living entities must be part of a system of relationships. They must interact with other members of the system to which they belong. A tiger in a zoo, though clearly a wild animal, cannot impart to us the full experience of wildness because it is not capable, in this setting, of interactive relationships. Indeed, the zoo setting cannot provide a system of relationships and nor, of course, could it ever be a system to which the tiger belongs. Even a family group of tigers is a mere single layer of inter-relationship. The idea of belonging was described in Chapter Four. Though not a rigid concept, it nevertheless carries with it the minimal requirement that it applies (for the wild animal) in the context of natural as opposed to artefactual communities.

Systems of relationship are important for experiencing wildness because it is only when these are in operation that wild entities can reach their full potential. That is, it is only when wild entities (animals or plants) are operating within a system of

---

1 Human artefact is regarded, here, as equivalent to human impact. For example, a clear-cut forest or an oil spill is artefactual.

2 There may, however, be a logical incoherence in describing the moon, or outer space, as wild if there is no sense in which these places could ever be tamed.

relationships that their adaptations and survival strategies can come into play. Life cycles, pollination, hunting strategies and so on are all aspects of wild survival and of wildness itself. In order for these things to operate, and to be understood and appreciated by us, systems of relationship must be in place. Part of the experience of wildness, then, is the experiencing of the myriad interactions between wild organisms.

Yet it is not just the bald fact of interrelationship that is important. The intricate yet seamless way in which organisms interrelate, in which they fit with – or have mastery of – their environment is relevant to the experience of wildness too. Cooper (1993, p. 342) believes that it is our own loss of fit with the environment that has resulted in our admiration for 'the wild animal's complete possession of its environment'. He articulates the point further:

> As a living symbol of possession of an environment, the animal represents the antithesis of the technological artificiality of our present condition. More often than not, it belongs in a small and simple community where each creature has a natural place and role; absorbed in its environment, it is incapable of that ironic distance which we are increasingly incapable of closing; it fits itself into the rhythms of its life and the seasons, instead of endeavouring as we do to level them out; and its world is one, unlike our own, where things and other creatures have a consolingly stable significance and value.

The way in which wild animals interact so capably with other wild beings and their environment (an ability we have largely lost) is a significant part of our experiencing of wildness.

Participatory scale Wildness can be found all around us, wherever we are. We can observe, for example, the wild poppy that appears through cracks in the city pavement, or the procession of ants across our backyard, the peregrine falcon nesting atop the office block, or the swifts that fly screeching up and down our busy streets. These are what we might call aspects or degrees of wildness. Usually, though, when we feel that we have had a fully wild experience we mean something more than this. So far, the ingredients, as it were, of wildness are suggested as: The presence of living organisms; systems of relationship; and a predominance of nature. If this were all that were required for wildness, the interaction between a ladybird and greenfly on an apple tree in our backyard would offer us a wild experience. On a certain level, we could indeed say that this was so, but only if our canvass for consideration was very small. Whilst backyards have (or are) pockets of wildness, these pockets are swamped, on a human scale, by the artefactual – the houses and streets that surround. The scale of such wildness is too small for us to feel a part of it; we can merely be observers. The factor that is essential if we are to experience wildness fully is, I suggest, the kind of scale that allows our participation in the wildness. The area must offer sufficient scope for our physical involvement – as well as for a specific kind of involvement.

To begin with, in order to experience wildness fully we should be able to *explore* the area. The canvass ought to be large enough for us to physically walk around and become a part of. We may explore and become lost or we may successfully navigate our way through. But even if we know the area well, it can still offer a degree of unpredictability, just as a regularly-visited badger sett provides us with a different experience every time. We should be able to feel that we are, or could be, participants in the experience. Wildness is something that impinges upon us, but also surrounds us, encompasses us. The apple tree may provide a sufficient arena for a greenfly to be wild, but not for us to fully experience wildness.

Yet it is not merely that an area must be sufficiently large for us to participate in its wildness; our experiencing of wildness is affected by the kind of participation we engage in. When we use our own wild skills to observe or interact with the wildlife, we tend to think that our experience of wildness is enhanced. We may use stealth to creep up on a grazing deer; we may employ camouflage to enable us to watch badgers; we may follow tracks and read other signs left by feeding animals; we may adopt body language to enable us to approach animals more closely; we may learn to identify plants and bird songs. To fully experience wildness, I suggest that we require a size of arena that enables meaningful participation, and the kind of participation that both utilises our own wild skills and heightens our awareness of the wildness around us.

Otherness Many argue that part of the experience of wildness is the experience of 'otherness'. Wildness is seen as the antithesis of those aspects of ourselves traditionally considered to differentiate us from other animals: formalised morality, artificial selection and at least some other kinds of artefact. Wildness is often viewed as an antidote to the human condition. As Cooper suggests, it may be the 'need for the 'other' which underlies the concern for the wild' (p. 340), while Birch (1990) believes that 'the essence of otherness is wildness' (p. 11). In the sense that wildness is 'other' there is no yardstick (except itself) with which to judge it. It cannot, for example, be regarded as unkempt, or overgrown, badly structured or poorly drained; a wild place is merely itself. As Livingston (1994) puts it:

> Wildness is not acquired through covenant or dispensation. Wildness is, and has been, from the beginning. It is not merely an evolved phenomenon; it is a quality of being, and a precondition of having become. As such, it is beyond the reach of rationality; it is previous, and transcendent. It has no missing parts, either through mutation or amputation. It requires no prosthetic devices, no fixing, no reordering, no moral overlays. Wildness requires no organizational intervention, even of the purest and highest democratic sort. Wildness is whole (p. 172).

The 'otherness' of wildness may have different effects upon different people. In certain situations, otherness may provoke fear (the feeling of being lost, for example). It may trigger a sense of challenge (the desire to 'conquer' a wild place by traversing or taming it). Alternatively, 'otherness' may produce surprise or delight, such as might be experienced simply by encountering a plant or animal in its natural habitat. And perhaps the fact that the 'otherness' of wildness puts it outside our usual framework of judgement and appraisal accounts, in part, for the sense of peace it brings many of us. If wildness just is then we may feel that we can simply *be*.

Part of the 'otherness' of wildness is to do with the notion – as described above – of unpredictability or surprise. Unlike so many aspects of culture, wildness is not under our control; wildlife, for example, cannot be 'conjured up' on request. We are merely one of the players in a wild place, not in charge or in control. But wildness is also 'other' in the sense that it is *predictable* in a way that we are not. We know, for example, that the swallows will amass prior to migration South in the autumn, that badger cubs will appear above ground in April or May, that the cinnabar moth caterpillar will feed upon ragwort in summer, and that (barring calamity) *this* whitethroat will return to *this* patch of scrub again next spring. These rhythms or cycles are no longer integral to our own way of being. This particular kind of predictability is not a strong feature of our lives today. Certainly, as animals, we are bound by our physiological makeup to behave in certain predictable ways, and culture, too, can engender great predictability. But we are not bound, in the way that wild animals and plants are, to this kind of *seasonal* predictability. These rhythmic cycles are driven by survival strategies. These in turn rest upon the complex inter-relationships between, for example, pollinators and the pollinated, those who eat and those who are eaten, or upon temperature requirements for, say, seed germination or ripening, hibernation or the birth of young. The human animal has, to a very large extent, cut itself free from such dependencies. Those for whom they still apply are regarded as 'other'. And this particular form of predictability is as responsible for our sense of wonder and awe about nature as the unpredictability of wildness is. We are enthralled by patterns and cycles in nature just as we are by the thrill of the chance encounter.

Nature's narrative An articulation of wildness would not be complete without considering one final aspect of it, namely, the requirement for a narrative that belongs to the 'other'. That is, we experience wildness in full when nature is telling its own story. Wildness is not merely a visual or aural phenomenon; part of the wildness experience is connected with process, inner dynamics, or what goes on 'behind the scenes'. For example, on visiting the Scottish Natural Heritage-owned island of Rum in northwest Scotland, the initial, visual impression might feasibly be one of wildness. There are magnificent peaks over which soar golden and white-tailed eagle, and upon which nest a large population of manx shearwater and roam groups of red deer. To all appearances, the situation is paradigmatic of wildness. However, all is not as it seems. The red deer, it turns out, are part of a scientific investigation that appears to be driven by questions pertaining to red deer management. Far from being freely grouping populations, the deer are artificially segregated, culled according to age or sex, and strictly monitored throughout the year, while some individuals wear plastic identification collars. On discovery of the story behind the scenes, the wildness experience is, at the very least, tainted. Clearly, there are varying degrees to which this might apply. But we can say that for an experience to be fully wild, there should not be some kind of human-dictated agenda behind the scenes, which perverts nature's narrative.

This is not to argue that human intervention *per se* will eclipse nature's narrative and compromise wildness. As we have already seen, human impact and wildness are perfectly compatible. However, human impact can be more or less compatible, more or less compromising. What is damaging to the wildness experience is when the human agenda is driven by a goal other than the preservation of *nature*. In the case of Rum, the red deer project is not underpinned by a desire to best-manage the deer as integral components of that natural system, but rather, by a desire to best-manage deer on hunting/shooting estates on the Scottish mainland. The deer on Rum, therefore, are living out a human-dictated narrative and not their own. To put it another way, the current deer management system on Rum is not inspired by 'nature in the raw'; it is not a programme that is primarily *for* nature. This is at the heart of the problem with Budiansky's (1995) line on nature. He argues for the extensive management of nature, but makes the mistake of assuming that, for example, because '[n]ature clear-cuts, so should we' (p. 72). His mistake is in his failure to recognise that the two *processes* are very different: one has a human agenda and tells a human story; the other tells nature's story. We would still regard a forest that had been devastated by a tornado as a wild place, whereas the wildness of a clear-cut forest has been compromised.

Wildness has its own narrative that we cannot reproduce because we are bound by our cultural narrative. Vicky Hearne (1987) articulates the difference between the two narratives. Writing about domesticated animals she argues that we inhabit a very different story in the case of domesticated animals as compared with wild animals. She builds up a rich picture to show that domesticated, but not wild animals can, through language, be brought into a mutually responsible, moral relationship with humans. Although it seems that 'communication' rather than 'language' would better describe the interplay between humans and domesticated animals, Hearne's point seems to be sound. Her central point is that our relationship with those animals that have been incorporated into human culture is different from our relationship with those outside of culture. She has come to this understanding through her experience as an animal trainer. The act of training that is fundamental, for she believes that when training dogs or horses:

> implicit as well as explicit in the trainer's language is the notion that animals are capable not only of activities requiring "IQ" – a rather arid conception – but also a complex and delicate (though not infallible) moral understanding that is so inextricably a function of their relationship with human beings that it may well be said to constitute those relationships (p. 8).

In contrast, if the author tries to train a wolf, no such moral understanding can be arrived at. Hearne explains that '[t]he wolf has wolfish social skills, but he has no human social skills, which is why we say that a wolf is a wild animal' (p. 22). Interestingly, she goes on to say that 'since human beings have for all practical purposes no wolfish social skills, the wolf regards the human being as a wild animal, and the wolf is correct'. For Hearne, wild and domesticated animals inhabit different stories, the latter having been incorporated into the human story. The wild animal that is kept captive is very clearly in the wrong story. Hearne illustrates this point with the story of Moja a chimp who had been raised by humans and taught sign language. Hearne recounts that she is appalled when Moja badly bit her handler's kneecap, despite knowing it to be wrong. This, she says, 'is not the sort of story I am accustomed to' (p. 34). Clearly, it is not possible to change the wild narrative simply by forcing wild animals to live with us. And talking later of another chimp, Washoe, Hearne recalls that Washoe's handler Roger Fouts, hoped and planned to turn her loose in Africa. She writes: 'I am mystified by this; I want to sputter something like

"But I thought you loved her!" ... I feel foolish ... it is clear that I am in the wrong story' (p. 38).

The point about narratives has been illustrated by considering single aspects of the wildness narrative – animals and plants – but the argument readily extends to the experiencing of wildness proper. The wildness of a place is in part determined by the story being told by its different elements, be these the animals and plants or indeed other elements such as rivers, mountains forests or deserts. A river that has been dammed, a mountain mined for tin, or a clear-cut forest, no longer tell their own stories, just as a domesticated animal no longer tells its own story, but ours.

I have considered six aspects of the phenomenological experience of wildness. None are adequate on their own, but together they seem to capture the essence of the wildness experience. In order to feel the wildness of a place we need to experience predominantly nature (as opposed to artefact). This must be organised into living, interrelated systems in which we might meaningfully participate, and about which there is a sense of 'otherness'. There needs, too, to be the sense that nature's story – and not ours – is being told. Having articulated what is meant by wildness, we need, now, to turn our attention to the second task. I earlier claimed that wildness is of central importance to conservation, indeed, that the overarching aim of conservation ought to be the perpetuation of wildness. It is now time to justify this claim.

Wildness as conservation goal

One claim that runs through the book is that the primary role of the conservationist is to safeguard nature such that it remains recognisably the nature we love and cherish. This is an uncontroversial claim. However, as I noted earlier, this very general aim needs to be translated into a more specific aim, because it is not immediately clear how nature can be best safeguarded. In Chapter One, I argued that the continued opportunity for speciation is a key factor in the safeguarding of nature, but this helps us little in determining how the conservationist ought to proceed. What is required is a tangible, attainable goal. That tangible, attainable goal, I contend, is the safeguarding of wildness. My reasons for making this claim are twofold. Firstly, other possible conservation goals are either implausible or inadequate. And secondly, by safeguarding wildness, we could not fail to safeguard the nature we so cherish. Indeed, as argued earlier, wildness is central to the very meaning of nature conservation.

To suggest that the aim of conservation is the protection of *nature* tells us very little about how to proceed because nature is to be found everywhere – between the cracks of pavements just as much as atop mountains or throughout forests. We need to know *which* nature – or at least what *kind* of nature – is worthy of protection. Central to conservation thinking is the notion of 'biodiversity': if we protect sufficient variety (particularly species and habitats) we will successfully protect nature. This notion directs the majority of conservation policy. The concept of biodiversity is clearly an important one: without variety the nature we cherish would cease to be. However, although I concede that biodiversity ought to remain a central conservation theme, there are good reasons for believing that it is a less helpful notion than 'wildness' as the means of safeguarding nature.

There are, in fact, a number of reasons for rejecting 'biodiversity' as the conservationists' means of safeguarding nature. The first is to do with difficulties surrounding biodiversity's main unit – the species – difficulties that will be articulated in Chapter Six. In short, I argue that the species is an arbitrarily defined unit, which makes it problematic as a means of determining biodiversity. And the fact that 'species' is a problematic category implies that 'biological diversity' is a problematic category too. Not only are there difficulties with the species concept, there are problems, too, in determining the boundaries of all holistic categories (such as a habitat).

Secondly, as unfolds in Chapter Eight there is no clear way of evaluating biodiversity. It is not clear when the demand for biodiversity is satisfied. For example, given that we wish to safeguard species *x* or habitat *y*, how many instances of *x* or *y* will be sufficient? It is unclear whether biodiversity ought to be assessed globally, continentally, nationally, or very locally. There are no clear guidelines.

My third reason for rejecting the promotion of biodiversity as a means of safeguarding nature is to do with the two biological categories 'the native' and 'the rare'. These categories, which are given conservation emphasis primarily because biological diversity is so highly valued, have themselves proved problematic as conservation values (see Chapters Two and Three). And finally, biodiversity cannot be considered sufficient as the means of safeguarding nature because it sanctions the keeping of (breeding) members of species in captivity as representatives or tokens of their species, even when there is nowhere left to release their offspring. That is, it allows nature to be safeguarded in a sterile, artificial, non-interactive, non-participatory manner that does not accord with the nature we know and cherish.

Bearing in mind the above difficulties with the concept of 'biodiversity' (as it is currently understood) and notwithstanding that most of the arguments are yet to come, I reject 'the promotion of biodiversity' as the conservationists' primary goal and means of safeguarding nature. This, then, leaves the way open for a more promising candidate. I propose that 'wildness' is such a candidate.

It is not merely that 'wildness' has credence as a candidate, as it were, by default. It is also an excellent candidate in its own right. To begin with, the idea that wildness is what matters in conservation resonates with the 'thinking interventionist' position of Chapter Four. With its requirement for both the predominance of nature over artefact and nature's narrative, wildness accords with the thinking interventionist's reference to raw nature. Wildness also escapes the problem associated with more concrete notions (such as nativity) that nature is thought of as if it were an ark. The notion of wildness is sufficiently indeterminate to allow for change, for the unfolding of a future nature.

Wildness gains further credibility as conservation's goal, from its acknowledgement (and substantiation) of the idea that conservation is very much a human enterprise, and as such, not bound by rigid, objective criteria. The notion of wildness is not concrete, measurable, or empirically determined in the way that nativity, say, purports to be. Since it is explicitly open to a degree of interpretation, it cannot be accused of being unrealistic or disingenuous. In other words, given that conservation is arguably as much about human feelings, passions and individual idiosyncracities (see Chapters One and Eight) as it is about concrete, measurable data, these aspects ought to be reflected in conservation's primary goal.

The wildness criterion fits comfortably with both the 'thinking interventionist' position and the nature of the conservation quest. It also accords with what was identified in Chapter One as a key factor in conservation: provision of the opportunity for natural speciation. The continuation of speciation is a vital criterion for ensuring that nature remains recognisably nature. As noted by Tickell (1995, p. 29) (his use of the term 'wilderness' corresponding to the use of 'wildness' in this context):

> Expanding these types of controlled wilderness areas would allow not only conservation but also evolution, effectively absent in our island nature reserves. And that is something that biodiversity conservation for the long term must surely encompass.

The wildness experience offers exactly the right ingredients with which speciation can take place: an abundance of nature, an absence of inappropriate human intervention and, above all, the requirement for 'participatory scale' that secures the space necessary for speciation. Decreasing habitat, as illustrated in Chapter One, is the greatest enemy of the speciation event.

But could it not be that wildness is compatible with loss of biodiversity? A place could remain wild even if a number of species were allowed to become extinct. Wildness surely does not depend on maximum biodiversity. I have already noted that there are problems with the concept of biodiversity, but more pertinently, even though there is no direct link between wildness and biodiversity, there is an important *indirect* link. By promoting wildness, we promote the means by which biodiversity comes about – that is, speciation – which, I reveal in Chapter Six, is of greater meaning to conservation than the mere enumeration of species.

Lastly, the criterion of wildness is credible as the goal of conservation because it seems to capture the difference between the nature we do and do not want to conserve. For example, a city park will not fulfil all the criteria for wildness and so is not of particular interest to conservation. Although nature may predominate (grass, flower-beds, shrubs and so on) these things will be unlikely to be fully 'other'. They will almost certainly be artificially selected 'garden' varieties and may be planted or pruned to resemble patterns relevant to culture. In short, they are more akin to artefact. Furthermore, a park does not exhibit *nature's* narrative, but rather, the human narrative: it displays nature formally for the benefit of humans. Conversely, whether or not they are ultimately managed, those areas we designate nature reserves or conservation areas fulfil all the criteria for wildness. Though nature conservation is inescapably a human enterprise, nevertheless, there is a sense in which nature reserves and conservation areas exhibit both nature's narrative and the 'other'. Indeed, their very designation is indicative of the fact that these places are *for* nature.[3]

Wildness, then, is an experiential phenomenon and there are six criteria that need to be fulfilled in order for wildness to be fully experienced. My arguments in support of the claim that the safeguarding of wildness ought to be the

---

3 Designated areas are *for* nature both in the sense of being 'for the purpose of' as well as in the sense of being 'pro' nature.

conservationist's overarching goal include the argument that this will secure the opportunity for speciation, so enabling nature to be retained in a recognisable (and valued) form. It may seem an obvious point, but it is worth noting that in seeking to safeguard wildness, we seek to ensure its *future*. That is, part of the conservation remit will be to ensure the *robustness* of wildness. An example will illustrate the point. At Hawsewater in the Lake District a pair of golden eagles hold on to a territory, in which, in 1996, they bred successfully for the first time since the 1980s. Their territory is hemmed in by hill-walkers on all sides and although the two birds have adapted to the disturbance, it was thought unlikely that any newcomer would tolerate this (personal communication with RSPB warden). It seems, then, that when one or other of the birds dies, the possibility of the golden eagle holding a breeding territory here again is lost. If it is accepted that this golden eagle breeding territory is an important element of the wildness of the area, then it is clear that this wildness is not robust. In fact, even though first appearances may suggest otherwise, it is precariously balanced. In this instance, the appropriate conservation action would have been to redirect or prohibit walkers from certain key areas. In this particular case, the conservation body concerned (the RSPB) lacked the requisite powers. The point is that to properly safeguard wildness, it is vital to have a thorough understanding of the system of relationships or inner workings of each situation.

A further point requiring clarification is this: if conservation ought to be concerned with the safeguarding of wildness, how much wildness ought there to be? That is, how much wildness is sufficient? As might be expected, there is no straightforward answer to this question but I shall attempt to formulate a response. Conservation, we have said, seeks to safeguard nature in a form that we both recognise and value. In order for nature to be perpetuated in this way, the perhaps single-most important factor is the opportunity for speciation. This is because speciation is necessary to replace those varieties lost through background extinction. It is also because of the current anthropogenic extinction event (see Chapter One) in which extinction is vastly outstripping speciation and does not look set to change. Wild places, I have argued, offer that which is necessary for speciation to take place: a predominance of nature, space for participation on a human scale and space for species to form part of a system of relationships. It follows, then, that the amount of wildness required is that which allows for sufficient speciation to occur. This in turn is determined by the amount of speciation needed to maintain nature in its current, recognisable and valued form. Clearly, no numerical guideline can be offered, because conservation, as we have discovered, is a human enterprise with a human-dictated agenda, and as such, is open to judgement and interpretation. Nevertheless, as a general guideline we might say that conservation requires sufficient wildness for *the possibility of a natural speciation event with respect to all extant species*. It would not be sufficient that speciation *per se* can occur. There are different kinds of speciation, one of which – sympatric speciation, which occurs in the same place as the parent species – is associated only with very small organisms such as insects. It would not be reasonable to suppose that nature would remain the nature we value if, say, insects were the sole animal organisms present. The full range of speciation events must be able to operate, which includes allopatric speciation, or speciation that occurs as a result of a geographical barrier. Only this kind of speciation can operate upon large organisms such as pandas, alligators and giraffes. Though it is not possible to be very exact about conservation's requirement for wildness, we can be sure that, since speciation is neither keeping up with the rate of extinction nor likely to replace extinct species in the future, given the current state of affairs, conservationists have not yet secured sufficient wildness.

One final point concerning wildness requires clarification. In this chapter, the emphasis has been upon characterising the full wildness experience, for it is wildness in its most complete sense that conservation ought to be aiming towards protecting and promoting. But there are, quite clearly, *degrees* of wildness; it is not an 'all or nothing' phenomenon. And nor is it the case that wildness in its most complete sense is the only wildness of value. Value will be commensurate with degree, and it will always follow that some wildness is better than none. Urban nature reserves, for instance, cannot count as fully wild since they are mere islands within a sea of human artefact. They can never, therefore, be truly 'other' or offer the scale necessary for human participation. However, they do have a great deal to offer in terms of experiential value, because they keep us in touch with the idea and value of wildness, and provide scope for limited speciation. They enable nature's narrative to prevail to some degree and they offer us the opportunity to become familiar with systems of relationship. Pond-dipping sessions for school children, often a popular activity in urban nature reserves, teach about life histories and relationships between water-dwelling organisms. The value of 'semi-wild' areas should not be under-estimated. Consider Mabey's (1995 b) description of an urban wasteland, where it is clear that nature has resurfaced. He writes that though the area around the Lea River Navigation is 'about as bizarre and artificial an ecosystem as you can find' (it was teaming with examples such as a coot's nest on a floating car seat) there is something about the place that conjures an image of wildness, something that feels quite different from the image projected by, say, a municipal park. The area was unplanned (even a bit chaotic); it had evolved. It was also a place in which humans could engage – albeit on a small scale – with nature, for people were picking wild fruit and nettles, taking short-cuts, picnicking, rambling. There was intense human involvement in the place: a utilitarian canal, naturalized plants and 'a derelict water purification works (now a small nature reserve) ... where some of the old stonework has been made into a mock megalithic henge'. Still, there was a sense of nature's narrative, given the 'Three-metre high self-seeded London planes growing out of the river wall ... naturalized parsley ... regenerating false acacias as thick as they are in the mediterranean ... a sprawling, luxuriant, feral grapevine ...' and two cormorants that flew over Clapton Mosque. Mabey's canal habitat has elements, or aspects of wildness that, though not combining to form wildness in its fullest sense, are, all the same, of value. They represent part of the value of the wildness that, I have argued, it ought to be the conservationists' primary aim to protect, indeed promote. If nature is to be maintained as the nature we currently cherish, then it is with the perpetuation of wildness that the conservationist ought to be most concerned.

Conclusion

Conservation best-achieves its primary goal of safeguarding nature by securing and promoting that which we call wildness. This is an experiential phenomenon, made up of a number of different aspects. Wildness is experienced in full when these aspects (criteria) are fulfilled, but it is not an 'all or nothing' phenomenon and may be experienced to different degrees. All degrees of wildness have value, but only when each one of its aspects are fulfilled can wildness be said to be successfully secured. It is only then that the opportunity for speciation – that process by which nature is perpetuated in its current, cherished, form – is safeguarded with certainty. Safeguarding wildness, I argue, is the best practical way of securing the future of the nature we wish to protect. And wildness itself is the very essence of the natural world that generates our desire to protect it in the first place. Wildness is both inspiration and practical solution.

PART III
A NEW EMPHASIS IN CONSERVATION

Chapter 6

The problem with species: reviewing units of importance

Introduction

The current human-caused extinction event gives cause for concern in that it threatens the nature we know and cherish. The threat is particularly serious because the replacement of species (speciation) is becoming increasingly less possible. In this climate, it is understandable that conservationists should direct their efforts largely towards species preservation, and this is indeed the emphasis of the majority of conservation projects. Even when emphasis is placed upon wider categories of living things such as habitats or ecosystems, this is often justified by a reference to the species that such systems sustain. It is, of course, difficult to talk of conservation without making reference to species, and indeed, in the earlier discussions of extinction, rarity and nativity, species were very much the focus of attention. However, in Chapter One I identified the key factor in the perpetuation of the nature we value, not as the preservation of species, but rather, as the opportunity for speciation. And in Chapter Five, I argued that this goal is best achieved through the perpetuation of wildness. So, whilst acknowledging the conservation position that species are important, I also suggest that they may not hold the central position in the overall conservation scheme that most conservationists assume them to have. But if species do not deserve to be central stage in the conservation arena, what position ought they to hold? In particular, if 'species' is of less importance than conservationists generally suppose, might there be more room for other categories of living thing whose relevance to conservation is currently overlooked? In seeking to assess the place 'species' has in nature conservation, we must begin by questioning whether – and if so, how – the category of species can be defined.

Problems with definitions

It is widely recognised amongst biologists that no single definition of species is quite satisfactory.[1] The oldest definition of species (and that used by the founder of taxonomy, Linnaeus) is the *morphological species concept*, in which species are distinguished from one another by their anatomical features. It is still the concept most widely used by taxonomists in the field. Indeed 'most of the species recognised by taxonomists have been designated as separate species based on morphological criteria' (Campbell, 1987, p. 41). As a definition, though, the morphospecies concept abounds with difficulties. It does not, for one thing, address the discontinuity that usually exists between species (that is, why species remain separate). Nor does it explain why two individuals that differ visually quite substantially – a male and female mallard duck, for instance – should be of the same species. As it is clearly so inadequate, there is no intention to discuss it further.

An alternative definition is the *biological species concept.* Mayr first enunciated this concept in 1942 and it is the definition adopted by most evolutionary biologists and population ecologists. Under this concept, a species is 'a population whose members are able to interbreed freely under natural conditions' (Wilson (1992), p. 36; emphasis removed). This definition does address the matter of discontinuity between species as it shows that reproductive barriers prevent genetic mixing. It also solves the puzzle of morphological differences within species because it explains that genes are shared. But, despite these improvements upon the morphological species concept, it is replete with its own problems. As Wilson, himself an advocate of the concept observes, '[t]his definition is an idea easily stated but filled with exceptions and difficulties ...' noting further, that 'not all biologists accept the biological-species concept as sound or as the pivotal unit on which the description of biological diversity can be based'.

The difficulties with the biological species concept – as discussed by Wilson himself – are wide-ranging. For example, there is the problem of what Wilson terms 'semispecies', 'populations that partially interbreed ... enough to produce a good many hybrids under natural conditions [a problem that is] acute in many plants' (pp. 42–43). Wilson dismisses the problem by noting that, in the case of interbreeding oak species '[h]ybridization between the species is much less frequent than breeding within them, and as a result the gene pools stay partly closed'. But this seems quite unsatisfactory. No more satisfactory is his observation that 'an abundance of hybrids ... may not be a global phenomenon in plants'(p. 43). The problem is that the existence of the phenomenon undermines the biological species concept.

---

1 Whilst recognising that there are a number of contemporary species concepts such as that based on phylogenetics, the two species concepts most widely employed in science and recognised by the population at large are chosen for consideration. Although no argument is offered here, alternative species concepts have been found to be no less problematic than the two dealt with here. For example, the Cladistic Species Concept is unsatisfactory because, amongst other problems, it cannot stand on its own. It defines a species as a group of organisms between two speciation events but relies upon the Biological Species Concept in order to define a speciation event.

The concept is further undermined by the existence of organisms that reproduce asexually. Again, Wilson acknowledges the problem in saying that '[t]he idea of the closed-gene pool has no meaning at all for the minority of organisms that are either obligatory hermaphroditic – those that have both ovaries and testes and fertilize themselves – or parthenogenetic, producing offspring from unfertilized eggs' (p. 44). Asexual organisms can, in fact, 'be assigned to species only by the grouping of clones that have the same morphology and biochemical characteristics' (Campbell, p. 462). But as we have seen, the morphospecies concept is seriously flawed. Wilson admits that, ultimately, 'the lines drawn by the biologist around such species must be arbitrary'. Nor can the biological species concept accommodate chronospecies 'which are stages in the evolution of the same species through time' (p. 44). That is, we cannot know whether any present species would be able to interbreed with one or other of its ancestors were it possible to bring them together. Perhaps Wilson is correct here to retort that '[t]he question is hollow when presented away from context', but still, there remains the problem that chronospecies are being identified by means of a defective theory.

Some may counter that the problems with the biological species concept are marginal ones. After all, as Wilson notes, despite hybridisation, species manage to remain distinct, and 'the vast majority of species are in fact sexual' (p. 45). Might it not be that the concept 'works well enough in enough studies on most kinds of organisms, most of the time'? It may be fair to argue this, but it would not be fair to give the impression that the problems with the biological species concept are marginal ones only. There are difficulties of a more pervasive kind, resulting in inconsistencies in everyday use of the concept. As Campbell observes, '[e]ven for populations that are sexual, contemporaneous, and geographically contiguous, there are cases where the biological species concept cannot be applied unambiguously' (p. 462).

Illustration of some problems with the biological species concept and its use in the way we categorise nature

Differentiating species and subspecies

The central problem with the biological species concept is the confusion that abounds between classification of a species and a subspecies. Cases of chain species, or ring species illustrate this problem. In one example, there are found, in the Rocky Mountains, four populations of the deer mouse (*Peromyscus maniculatus*). Since the populations are phenotypically distinct and separated geographically, some experts designate them as subspecies. But because they generally interbreed at the overlap between the populations they are considered to be the same species. The difficulty arises because two of the populations do not interbreed. Since their gene pools are not completely isolated (due to the chain effect) and it is possible that genes could flow from one to the other non-interbreeding population, some feel it is legitimate to regard them as the same species. However, '[t]he extent of gene flow by this circuitous route is probably so slight that even the experts disagree on whether or not these two populations should be called subspecies or be assigned to separate species' (Campbell, p. 462). This might not be a great problem, but for the fact that, as Wilson himself admits, '[t]he uncertainty of the limits of populations combined with the discordance of traits means that the subspecies is an *arbitrary unit of classification*' (p. 60; emphasis added). If the subspecies is an arbitrary grouping within a species, and there exists confusion over whether certain of these might actually be species, then it casts serious doubt on the usefulness to us of the biological species concept as a means of distinguishing species. It also casts serious doubt on the viability or integrity of the concept itself. The confusion that exists over which groups to classify as species and which as subspecies can be illustrated by an example closer to home. There is much disagreement over whether the Scottish crossbill is a subspecies of the crossbill or whether it constitutes a separate species. It seems that the Scottish crossbill is so similar to the crossbill as to make identification in the field – except in the Scottish crossbill's native Caledonian relict forest – uncertain. But if this is the case, then it is unlikely that we can be sure the two types are not interbreeding at the edges of their respective habitats. The Scottish crossbill may merely be a subspecies. Indeed, given that geographic location is seen as the only sure way of identifying it, this suggests that the Scottish crossbill can only truly have subspecies status.

Limitations of the biological species concept

This exposes a further general problem with the biological species concept. According to this concept, the delineation of a species does not require that the group look different to other groups of organism, merely that they are reproductively isolated. However, the only way that reproductively isolated groups can be *recognised* is if they look (and behave) differently. If there is a particular reason to look closely, one often finds that there are more species than supposed, these being recognised by visible differences. Wilson, for instance, describes how, in efforts to understand the spread of malaria, it was discovered that the mosquito *Anopheles maculipennis* is actually a group of at least seven species and not, as was supposed, a single species. Close inspection revealed distinctive characteristics such as 'the size and shape of the rafts of eggs laid by female mosquitoes on the water surface' (p. 41). It is true, of course, that species, as a matter of fact, do usually look (and behave) differently from one another. But my point is that the morphospecies concept is often required in order to recognise, in the first place, the existence of reproductively isolated populations. That is, the morphospecies concept is frequently required in order for the biological species concept to be usable.[2] This may not be seen as a problem in that the former concept backs up the latter. However, it does mean that the delineation of species is limited by our ability to observe well in the field, by our desire to look more closely, and by the sorts of

---

2 It is true that the isolation of a population may also be a factor in alerting us to the existence of a species, but in the absence of obvious isolation, we rely upon morphological and behavioural differences (which are usually linked). It is also true that species are sometimes distinguished by biochemical tests or mating experiments (Wilson, 1992; p.42) but these are the exception and not the rule.

difference that make sense to us. This illustrates further, the inadequacy of the biological species concept as a means of identifying species.[3]

Problems of application: muddled thinking

If the biological species concept is inadequate as a tool for identifying species, our habit of applying it haphazardly creates additional confusion. As we have seen, under this concept the definition of a species is a population whose members interbreed with each other under natural conditions. The understanding is that the members breed only with each other and not with the members of other species. Although we seem to go to great efforts at times to determine whether groups comply with this requirement (for example through mating experiments) at other times, we ignore the requirement altogether. Consider, for instance, the case of red deer and sika deer. They interbreed readily and the hybrids are very successful, yet the two types of deer are considered to be distinct species. Indeed, their propensity to interbreed (and hybridise) is of grave concern to conservationists. There is clearly something of a muddle. A similar story can be told about the red wolf and the coyote in the USA. With the red wolf numbers dwindling, this species began to interbreed with the coyote. Desperate to save the red wolf species, conservationists employed very drastic measures to discourage interbreeding. This included an attempt to create a coyote-free buffer zone by setting leg-hold traps and shooting (De Blieu, pp. 39–40). As discussed in Chapter Three, in Britain there is, periodically, quite a furore over the RSPB's shooting of ruddy ducks, a species that has begun to interbreed (and hybridise) with the white-headed duck in Spain. The 'purity' of the (rarer) white-headed duck is prized over the lives of members of the ruddy duck species. But with what definition of 'species' are conservationists working? It is more curious still, when we consider the case of the carrion crow and the hooded crow. These two crows, reckoned to be subspecies of the European crow, are regarded as the same species because they interbreed freely where their ranges overlap, their offspring being fertile. However, it seems that the zone where interbreeding occurs is narrow. This is an indication that the hybrids are less successful breeders than either of what might be called the 'pure' forms. What is odd, is that these two crows, whose offspring (though fertile) are clearly not particularly successful, should be assigned subspecies status, when other groups of animals (such as the red deer and sika deer) are considered to be separate species despite the success and spread of their offspring. Taxonomists and conservationists alike are insisting that species both can and ought to be defined, while at the same time failing to do so. Their characterisation of 'species' is singularly arbitrary. Wilson, who is clearly a supporter of the biological species concept characterises those plants that tend to hybridise as 'semispecies' (p. 42). For the sake of consistency, a number of groups of animals currently thought of as species

---

3 A new classification system – using DNA – has been given some airing in the press. Whether such a system will become popular or not, it clearly would be troubled by this particular problem of the Biological Species Concept.

(including those mentioned above) ought, presumably, at the very least, to be re-classified as semispecies.

Species-status and domesticated animals

Perhaps, though, it is the status of domesticated animals that is subject to the greatest inconsistency. There is no agreement, for instance, on whether they should by assigned species status or not. Linnaeus' *Systema Naturae* (1758), 'internationally accepted as the basis for zoological nomenclature' includes all the common domestic animals. However, as Clutton-Brock (1987) observes, it is recognised today, 'that there are a number of inconsistencies and difficulties that make the application of the Linnaean system of binomials to domestic animals rather awkward' (p. 194). This, she says, 'is because our ideas of what is a species have changed and become more complicated'. We cannot, now, for instance, assign the dog and the wolf to different species simply because they look different, as Linnaeus clearly did. It is because of such difficulties that, according to Clutton-Brock, 'it has been suggested that domestic mammals should be excluded from formal zoological nomenclature [though] up to the present no agreement has been reached on what system should take its place'.

Generally, domesticated animals are described as breeds, this category being considered as distinct from subspecies. Whereas a subspecies evolved as a result of reproductive isolation in a restricted geographic locality, a breed is merely a product of artificial selection in which geographical barriers are irrelevant. The problem is, there are groups of animals that fall somewhere between these two definitions. One example, given by Clutton-Brock, is the Soay sheep. This is a population that:

> has undergone little artificial selection since it was first taken to St. Kilda, probably during the Neolithic period, but has evolved into a uniform group adapted for survival in the tough conditions provided by a small island in the North Atlantic (p.29).

It is not at all clear that this animal is any different in status to, say, the various subspecies of field vole thought to have reached their respective Scottish islands (such as Muck and Arran) by human introduction (Yalden, 1982, p. 35). Can we be sure, for example, that the Soay sheep underwent any artificial selection other than the fact that the original population was chosen by humans? Is this significantly different from the accidental transfer of animals such as the field vole and common shrew?

It is odd that although domesticated animals are not to be considered species or even subspecies, they can still be thought to form *clines*. Within a species, different populations often merge into each other and 'when this occurs across separate geographical regions the subspecies are said to form a cline' (Clutton-Brock, p. 30). Though not a taxonomic unit, a cline is 'a gradation of measurable characters that are unidirectional and pass into each other without discrete breaks in sequence'. The discordant factor is that clines 'develop as adaptations to local environments and micro-climates' (such as the regional distribution of coat colours and markings in the domestic cat (p. 32)), yet domesticated animals are denied subspecies status on the basis that they are the product of artificial selection by man and do not constitute discrete geographical units. It seems quite clear, though, that if domesticated animals can form clines, they at least partially fall under the definition of subspecies. There are certainly no clear-cut distinctions to be made.

Underlying the difficulty of which category to assign domesticated animals to is the question of whether or not they 'should be treated as taxonomically identical with their wild progenitors' (p. 195). In other words, should they be assigned to the same species? The domesticated Indian elephant is given the same name as its wild counterpart because, as Clutton-Brock says, 'they will both belong to the same breeding population' (p. 194). However, since man closely controls the reproduction of most other domesticated animals – meaning that wild and domesticated populations are effectively separated – there is seen to be a problem in the different populations sharing a name. But it seems strange to worry about assigning two populations to the same species because they are artificially prevented from interbreeding. They can potentially interbreed and indeed frequently do, as the numerous feral populations bear testimony to. There is no reluctance to name free-living, interbreeding populations as separate species (such as red and sika deer or coyote and wolf).[4] Conversely, if populations of animals are to be assigned to separate species merely on account of their differing morphology or behaviour, then it seems quite reasonable to consider the dog to be a different species from the wolf. Coppinger and Schneider (1995), in discussion about the evolution of the dog, suggest that the real difference between the dog and the wolf is not anatomical but behavioural, for 'dogs are tameable ... and trainable, wild canids are not' (p. 35). Indeed, there is more morphological variation, they say, between the dog breeds than between the rest of the canids (p. 33), yet dog behaviour, which is very different from other canids, is more or less standard across the numerous breeds.

If we were to stay true to the biological species concept, while accepting domesticated animals into the taxonomic system, then it is clear that domesticated animals must belong to the same species as their wild counterparts. Although the two populations are often artificially held apart, they can (and do) interbreed. The dog, for example, might be considered a subspecies of the wolf. If, however, we deviate from the biological species concept, we can quite easily envisage the dog as – anatomically and/or behaviourally – distinct from the wolf and as a separate species. Alternatively, we could exclude all domesticated animals from the taxonomic system or devise (and justify) a separate category for them. Our present system follows none of the above forms and represents nothing less than a complete muddle.

Taking the dog again as an example, it seems that it has been allocated to a separate category on the basis of its artificial history, its different forms (for example, the Border Collie and the Poodle) considered to be breeds rather than subspecies. But the justification for this is unclear. Following the biological species concept, the dog is the same species as the wolf (assuming the latter to be its

---

4 The anomalies in the current taxonomic system are compounded by the fact that three of the domesticated mammals discussed in Clutton-Brock's book have the same name as the wild species: the rabbit, the reindeer and the Bali cattle, this being 'because there is no tradition of naming them separately' (p.195).

ancestor). Further confusing the picture, is the fact that the dog is still, taxonomically, a separate species, albeit due to Linnaeus' mistake. The result of the current confusion is that authors, aware of the Linnaean mistake, and so reluctant to describe the dog as a separate species, find that they are unable to categorise the dog as anything at all. It is not fitting to call *the dog* a breed, because breeds are varieties of dog. 'Variety', 'race' and 'strain' are even less helpful, because they describe diversity within populations *below* the level of subspecies and breed (Clutton-Brock; p. 31). We are left wondering, then, what sort of description is thought to fit 'the dog'? The authors Coppinger and Schneider (1995) are happy to classify the dog as a species since they believe 'the ancestor of breeds of dog was the dog itself' (p. 35; see also, Coppinger & Coppinger, 2001)). Most authors, however, do not take this view and are left wondering exactly what to say about the dog. Clutton-Brock, for instance, in discussing the way in which the domestication of the wolf to the dog took place, suggests that individual animals exhibiting certain character traits would have been given preferential treatment. She then explains that '[t]hese characters were inherited and so in time a separate kind of animal evolved, the *dog*' (p. 38). But what kind of animal is it?

Defining 'species' is far from straight forward. The two most widely used definitions – the morphological and the biological species concepts – have many problems. These inherent problems are compounded by our haphazard use of the terms. The question that now arises is whether our definitions of 'species' are at fault or whether species are simply not definable.

Are species real?

It is quite clear that no one definition of 'species' is adequate. Although a supporter of the biological species concept, Wilson (1992, p. 46) recognises the limitations of any single concept:

> Someday biologists may come up with a single concept that unifies sexual species, nonsexual species, and chronospecies into a single, theoretically powerful natural unit. But I doubt it. The dynamism of the evolutionary process and the individuality of species make it unlikely that a completely universal species definition will ever be fashioned. Instead, two to several concepts will continue to be recognised, like the waves and particles of physics, as optimal in different circumstances.

So if species are definable, it appears that they can only be so in the above sense. Does this suggest, then, that species are not real? The answer to this depends on what is meant by real. Species are real in the sense that there is something in the world that suggests the concept. That is, there are real populations of (loosely) reproductively isolated animals and plants. There are also real populations of morphologically and behaviourally similar individuals, and populations that share genes and a common heritage.[5] The concept of species is rooted in the real world; it is not *merely* an idea.

Nevertheless, the sense in which species can be said to be real is limited. The *concept* of 'species' (like any other concept) is an abstraction from the real world. In other words, it is only a partial description of the way things are; one 'appearance' of it that relates to a particular context. There is nothing in the world that *necessitates* this kind of categorisation. It is merely *a* way of describing the natural world and not *the* way. This is not, however, to argue that any way of describing the natural world is as meaningful as any other, merely that there is no single way of doing so. The reality of species is contingent upon our particular way of viewing the world. 'Species', then, is just one of the abstractions we use to label what we find in the real world of things such as individual animals and plants. The suggestion that 'species' is an abstraction is supported by the argument that species are not natural kinds. Such a view is taken by Clarke (1994) and by Ingold (1995). The latter states that:

> ... [b]iological taxa are not natural kinds, so that they cannot be expected to conform to any essentialist system of classification. Indeed, most biologists reject any appeal to essentialism in the definition of species, as fundamentally contrary to the Darwinian principles that lie at the core of their science (p. 15).

In other words, there are no necessary and sufficient conditions for being, say, a tiger. It would seem, then, that the natural world is not, contrary to popular opinion, *made up* of species. Rather, 'species' is an abstraction that we have imposed upon the natural world. Insofar as there are real populations whose individuals have in common certain features (such as morphology or breeding potential) it is perfectly reasonable to label these as species. However, while it is legitimate to do this (to work, as it were, from the bottom up), it is not legitimate to start with the concept of species and apply it universally (that is, to work from the top down). We are mistaken if we assume that the natural world will fit into either an existing or a future concept of species. There is no evidence of universally applicable natural boundaries of this sort.

If there are indeed no universal natural boundaries that correlate with the concept of species, is one taxonomic system of classification as legitimate as the next? Certainly, the Western taxonomic system has varied through time. As Sebeok (1994) notes, Aristotle classified whales as fish (p. 68) and, as already discussed, changes in thought about species have led to a confused situation regarding the status of domesticated animals.

Sebeok argues that taxonomic classification is wholly a culturally derived system. He claims that even the Linnaean plant/animal dichotomy may be variously represented, and that 'the choice of a classification scheme is ultimately (although,

---

5 Clarke's use of Kant's Realgattungen (real species) and Arten (artificial species), which distinguish between genuinely interbreeding stocks and morphological species, sets up a false dichotomy (Clarke, 1995; p.22).

of course, within limits) a personal matter' (p. 64). He goes on to say that taxonomy is: 'But a segment of the far more venerable as well as unbounded science of systematics which has diversity as its subject matter ... [depending] on a whole variety of alternative, presumably complementary, approaches' (p. 67). Plants, for example, 'could be reclassified according to how they fabricate their photosynthetic pigments'. Sebeok's main point is that '[a]ll organisms pertain at once to a plurality of codes, each of which is capable of being transmuted into every other'. Given this, categorisation of the natural world into species is merely one way of understanding it. Sebeok illustrates his point first with the observation that 'most laymen still believe that whales are, indeed, fish', and then with the follow-up comment that whales, as well as being mammals, are also fish 'and other entities – as Moby Dick – to boot' (p. 68).

While Sebeok's position has certain appeal, it is not fully convincing. It does not follow from the argument that there are no natural boundaries constituting species that any categorisation is as good as any other. The category 'species' is, nevertheless, determined by real differences and similarities in the world. And as Sebeok himself acknowledges, there are *limits* to the possible comparisons we can make, and so limits to the possible classification system we endorse. In practice, classification systems are remarkably similar, even where there has been no communication of ideas between the classifiers. Wilson (1992), for instance, relates how Ernst Mayr found that, on a visit to the remote Arfak Mountains of Papua New Guinea, the native hunters' classification of birds correlated almost exactly with that of his own (pp. 38–39). The hunters had a need to distinguish between the birds as they depended on them for meat, and the obvious way to do this is by way of morphological and behavioural differences. As I noted earlier, there is a fairly strong relationship between the biological species concept and morphology, so it is clear that differences detected visually will often happen to distinguish biological species. The important point is that humans organise or systematise things in remarkably similar ways because our makeup determines that we understand the world in particular ways. Our phenomenological experience of the world is rendered meaningful in fairly standard ways. What will make the difference between classification systems is firstly, incorrect information – whales are not, in fact, fish. Secondly, lack of motivation to observe closely is crucial. On Wilson's own trip to Papua New Guinea to collect ants he 'repeated the cross-cultural test and found that the Saruwaget people could not tell one ant from another'. It was not, he explains, 'that Saruwaget ants and natives failed the test, only that Papuans have no practical need to classify ants' (p. 39). And thirdly, differences are produced by a practical need for emphasising some differences more than others. One African tribe, for instance, while classifying snakes in the same way as Western science classifies species, classify their breeds of cattle in this way too. This in no way demonstrates a different system of classification but merely represents the importance that cattle have for this tribe. Indeed, it differs little from the Western classification system with respect to the muddle that currently prevails with regard to the classification of domesticated animals.

My response to the question about the reality of species, then, is to say that there are real similarities and differences in the world that we are so constituted to notice and label, and with which we understand the world around us. With respect to the natural world, these differences and similarities are grouped into something we have labelled 'species'. But the label 'species' is an abstraction. So although it is fair for us to say that the grouping of 'species' is meaningful to us, we cannot claim that the natural world is made up of these groupings. There are no natural species boundaries. To see species in this way is to recognise their importance as conservation labels; species-talk is universally recognised and understood. But it is also to acknowledge that since they are simply labels, they do not automatically merit the centrally important position they currently enjoy in the field of nature conservation. (I will look, shortly, at the question of how this argument fits with the earlier claim that speciation is of paramount importance to the safeguarding of nature.) This is not to say that the current system of classification is inadequate and that we should seek to establish new labels within alternative systems. It is merely to point out that there is no obvious reason why the category of 'species' is given preferential treatment over other categories within the current classification system. Other categories (such as 'subspecies') are equally contrived but useful labels. And this is not merely an academic point. There are real dangers, for conservation, in overvaluing the species.

Re-evaluating the importance of species: the danger of overvaluing species

Species as descriptive not prescriptive

One thing that the discussion so far has revealed is that 'species', as an abstraction from the real world, as a legitimate but convenient label for our use, is a *descriptive* term. But there is no justification for applying prescriptive terminology to the category of species. Although our label of 'species' is generally useful, there are no universal natural boundaries into which the category fits. Consequently, there are many examples of categories overlapping and appearing indistinct. Clarke (1994, p. 22) points out that the concept of 'palaeospecies' (or chronospecies) brings 'to our attention the fact that even the most rigorously xenophobic of contemporary species are, once we bring their past to mind, merely *Rassen* (varieties) of a single *Formenkreis*, or ring-species'. If we take the species concept too literally and too seriously, we are in danger of unjustified intolerance of the very natural process of hybridisation. This intolerance is very much in evidence amongst conservationists as was discussed in Chapter Three. How these reproductive matches can be regarded as hybridisation is the great mystery, because, as I have shown, under the best-used (Biological) Species Concept, in order to breed successfully, these animal pairs must be of the same species. There is, however, an almost fanatical concern with and desire to prevent such interbreeding. Such fanatical concern would be much better placed if directed towards the real threat to nature: species losses through habitat destruction.

The 'purity' of species, which seems to be of great concern to many conservationists not only smacks of biological fascism, it is intellectually misguided too. Hybridisation, as we can learn from Haldane (1932), is a perfectly respectable evolutionary strategy. It is, he suggests, a 'possible mode of making rapid evolutionary jumps ... [which] may lead to the immediate formation of a new species by allopolyploidy' (p. 105). He writes:

> Lotsy (1916), in particular, had emphasized the importance of hybridization in evolution, and shown that it occurs in nature ... [and that] it is difficult to doubt that hybridization has rendered possible the coming together of certain combinations of genes which could not have arisen otherwise (pp. 106–107).

If we use the species concept, then we must use it as a means to communicate and inform, rather than as a concept with real or known boundaries or prescriptive powers.

Whilst the species concept can be a useful tool in the fight against the loss of biological diversity, it is important to remain aware of its limitations. In particular, it is important to be sensible to the absence of definite boundaries between species and other categories (such as subspecies) and the slide between them. Not only should we be aware of the problem with emphasising species purity. So too ought we to be wary of the automatic prioritisation of species over, for example, subspecies, and the even more prevalent prioritisation of species over individuals. As Darwin (1985 (1859)) himself observed:

> Certainly no clear line of demarcation has as yet been drawn between species and sub-species – that is, the forms which in the opinion of some naturalists come very near to, but do not quite arrive at the rank of species; or, again, between sub-species and well-marked varieties, or between lesser varieties and individual differences ...
>
> From these remarks it will be seen that I look at the term species, as one arbitrarily given for the sake of convenience to a set of individuals closely resembling each other, and that it does not essentially differ from the term variety, which is given to less distinct and more fluctuating forms (pp. 107–108).

A misunderstanding of the species concept can result in species being held in too high esteem, such that their protection is advocated regardless of the cost to individuals, subspecies, or the wider environment. The problem is, of course, most likely to arise in the case of rare species. Conservationists must be careful, then, not to fall into the trap of assuming the environmental superiority of species over other (equally loosely-defined) conceptual categories, as well as (clearly definable) individuals. And, as argued in Chapter Four, there are good reasons to value individually bounded organisms above abstract categories of living thing such as species or subspecies. It ought not to be the case that species protection is to be attempted at *any* price.

The danger of undervaluing process

Another kind of danger for conservationists who fail to appreciate the true nature of the species concept is the likelihood of the product (species) masking the process (speciation). The former is often given greater attention than the latter. It does seem, however, to be a mistake to assume that biological species are all the same kind of thing because the various speciation processes seem to be qualitatively different.

There are differences, for example, in the degree of isolating change necessary for the formation of a new species. Wilson (1992) tells of the very small segregation between species of leafroller moths in the family Tortricidae. He says that '[t]he variation they manifest from one species to the next is the kind commonly based on mutations in only one gene' (p. 65). Variation might even be more elementary still, for some leafroller species are separated by:

> a difference not in the organic structure of the substances composing the sex attractants but in the percentages of the substances, in this case a variety of acetates, that go into the blend.

Once species are reproductively isolated, the genetic distance between them begins to widen. So the category of 'species' covers variety in the degree of isolating change that was necessary for their formation, as well as differences in genetic distance from parent species depending upon the length of time since speciation took place.

There are also differences in the nature of the isolating mechanisms. Some speciation is allopatric (in a different place) while other speciation is sympatric (in the same place). In the latter case, speciation may be through polyploidy (an increase in the number of chromosomes) or non-polyploid sympatric speciation such as 'through the intermediate host races of insects that feed on plants' (Wilson, p. 66). Sympatric speciation is to be widely found in the plant kingdom but in the animal kingdom it is mainly restricted to insects. The fact that sympatric speciation takes place on so small a stage means that when speciation takes place, the new species are genetically much closer to the parent species than is the case in allopatric speciation.

Finally, there are also significant differences in the time required for the formation of new species. At one end of the spectrum is allopatric speciation while at the other end is sympatric speciation. The former kind of speciation may take a very long time (many generations) whereas polyploid speciation 'is virtually instantaneous in its effect, potentially isolating a group of individuals from its ancestors in one generation' (p. 63).

Wilson captures the significance of these differences in the speciation process in his characterisation of species that can evolve rapidly, in large numbers, and with little genetic variation, as cheap species (p. 67). He writes that if such species fill a new niche:

> they probably do so with relative inefficiency. They have not yet been fine-tuned by the vast number of mutations and episodes of natural selection needed to insert them solidly into the community of organisms into which they were born ... Great biological diversity takes long stretches of geological time and the accumulation of large reservoirs of unique genes. The richest ecosystems build slowly, over millions of years ... A panda or a sequoia represents a magnitude of evolution that comes along only rarely. It takes a stroke of luck and a long period of probing, experimentation, and failure (p. 68).

Once we take into consideration the speciation process, we see that the term 'species' encompasses more variety than might be supposed. The different processes are suggestive of a qualitative difference that may easily be missed by conservationists putting too much emphasis on the 'product'. If what we are seeking to do in conserving nature in something like its current form is conserving the levels of variety we currently have, then the process of speciation – the way species have evolved – is of central importance to this aim.

Emphasis of the species concept not only masks differences in the speciation process, it also masks the process itself. The species concept is merely a way of expressing biological diversity. If conservationists fail to appreciate this then they are in danger of failing to value, sufficiently, the means by which diversity comes about and is maintained. And it is the means of maintaining future diversity that the conservationist needs to be concerned about. 'Speciation', and not 'species' is the important player in the perpetuation of diversity.

Bradley Wilson (1996) endorses this prioritisation in as much as he believes that the species is not an important unit of evolution. His argument rests upon the likelihood that one or other of the gradualist or the punctuated equilibrium view of evolution is correct, and the understanding that 'if species are units of evolution, they must be reasonably discrete temporal entities that have the capacity to evolve' (p. 409). Taking the gradualist view first, (the idea that species have evolved gradually by a process of tiny incremental steps), Wilson shows that this understanding of evolution is incompatible with the existence of species as discrete units over time. That is, 'the gradualness of the evolutionary process ... makes a division of organisms into species arbitrary' (p. 412). It is not, as Wilson explains, that 'gradualism makes it impossible to divide organisms into relatively distinct species relatively clearly' (for, as he notes, an essentialist view of species is not incompatible with precise delimitations). Rather, 'there is an indefinite number of equally satisfactory ways of doing so' (p. 419). Wilson rejects the possibility that a shift to population thinking solves the problem, such that 'the beginning of a new species can be identified with the geographical isolation of a founder population'. He does so, firstly because such identification can only be retrospective, and secondly, because any founder population 'is still reproductively and genetically continuous with the original species' (p. 413). Furthermore, a geographic isolating event changes only factors external to the organisms 'and does not create, by itself, any intrinsic biological differences between the two populations'. In themselves, then, geographic isolating events cannot serve to distinguish populations as distinct species.

The alternative (popular) view of evolution is Eldredge and Gould's (1972) theory of punctuated equilibria. This theory holds that there are long periods of stability or stasis in any lineage, punctuated by rapid speciation events following geographic isolation. As Wilson puts it:

> [t]he beginnings and endings of a species, it might then be argued, are marked by periods of rapid evolutionary change. Between these periods, species are maintained by stabilizing selection coupled with gene flow through the populations of the species (p. 414).

It would seem, then, that species are discrete entities and may be important units of evolution. However, according to Wilson, this does not follow, for '[o]n the theory of punctuated equilibria, species would seem to be, not units of evolution, but units of *stasis*' (emphasis added). Even if one were to identify a species with a lineage of populations, the species would not play an important role in the evolutionary process. Rather, the isolated population, which evolves by natural selection, would be the key player.

Bradley Wilson's argument that the species is not an important unit of evolution supports my claim that species are given too much emphasis by those seeking to safeguard nature (where this is understood to mean the *continuation* of nature). Wilson does not claim that species categorisation is either useless or misplaced. Indeed, he acknowledges the role that 'species' had in making accessible to others, Darwin's ideas, and its current use as a 'reference point for the purposes of studying the effects of natural selection over time' (p. 418). He acknowledges, too, the usefulness of its most popular manifestation – the Biological Species Concept – as 'a biological characteristic that is particularly relevant to the process of evolution, namely, reproductive relationships within and between populations of organisms' (p. 416). But, he says, it can be understood only as a 'synchronic species concept'. There are problems with its diachronic use (that is, its application to groups of populations over time). This is partly because of the 'practical difficulties in applying the criterion of reproductive isolation to temporally distant populations'. But mainly it is because:

> the fact that evolution produces certain patterns among contemporaneous groups of organisms, identifiable on the basis of reproductive relations, does not imply that those same reproductive relations, extended over time, pick out important evolutionary units (p. 416).

Use of the category 'species' will no doubt continue as a way of conveying ideas about the natural world and in this its use is both pragmatic and legitimate. Earlier discussions – about extinction, rarity, nativity and so on – would have been difficult without reference to species. However, the category 'species' should not be regarded as anything more than a descriptive aid. Certainly, it is important that those involved in conserving nature should not suppose the species to be the important player in the job of safeguarding nature that it is currently assumed to be.

There may appear to be a contradiction between the claim that the category 'species' is both problematic and of less relevance to conservation than supposed, and the claim that speciation is of paramount importance to conservation. Is not, for example, the concept 'species' implied by 'speciation'?

On one level, of course, speciation does not make sense unless species are assumed, because species are the end product, as it were, of speciation. But in another sense, speciation merely describes *the process that produces variety*, where that variety is most usefully – but in no way rigidly – delineated by 'species'. Biological variety or diversity may not be able to be pinned down – either by the category 'species' or indeed by any other category – but this does not make such variety any the less important. What it does do is to shift the focus of concern away from the end-product and towards the process that produces this variety – namely, the process of speciation. At first glance, then, there may have appeared to be a contradiction between the de-emphasis of 'species' and the promotion of 'speciation' but this worry is unfounded.

If one of the dangers of overvaluing the species is undervaluing the process, then we ought to be clearer as to what, exactly, this process that begets biological diversity is, and how it is to be valued. Various isolating mechanisms lie beneath the formation of diverse biological forms. The great majority of these are prezygotic barriers that 'impede mating between species or hinder fertilization of ova should members of different species attempt to mate' (Campbell, 1987, p. 463). Prezygotic isolating mechanisms include ecological, temporal, behavioural, mechanical and gametic isolation. There are postzygotic barriers, too, that prevent hybridisation between some species. But what is significant is that the great majority of these isolating mechanisms are expressions of differential adaptations to the environment.

To illustrate this let us look at an example used by E. O. Wilson: the giant silkworm moths of North America. They 'fly and mate at various times during late afternoon and through the night' the females releasing a powerful chemical scent to attract the males. Each of the 69 species, Wilson explains, 'has its own *heure d'amour*' and '[f]ew if any mistakes are made thanks to the combination of timing and scent' (p. 51–52). It is this adaptive isolation that prevents the majority of species from hybridising, and leads Wilson to say that '[t]he wonder is not that hybridization fails but that it ever works' (p. 50). The significant point is that it is the particular kind of interaction between organisms that brings about and sustains biological diversity. As Wilson puts it:

> the differences between species ordinarily originate as traits that adapt them to the environment, not as devices for reproductive isolation. The adaptations may also serve as intrinsic isolating mechanisms, but the result is accidental. Speciation is a by-product of vertical evolution (p. 53, emphasis removed).

If, then, we are to appreciate biological diversity meaningfully, we must appreciate, above all, the myriad of different interactions in the natural world. To do this is to pay attention to relationships between wild-living individuals and their environment. It is not enough that we have reproductive isolation. A zoo full of different, reproductively isolated species is a barren and artificial place, and does not capture the importance of biological diversity. What we are missing is an appreciation of the place that living organisms have in the environment, and it is relationship that lies at the heart of this. Relationship is central to an organism's adaptations, which in turn are the source from which biological diversity springs. It follows, then, that if conservationists are to avoid the mistake of valuing the product (species) over the process (speciation) they must pay more attention to the primacy of relationship.

The danger of neglecting individual relationship

Over-valuation of the product of evolution carries with it the danger of neglecting the process of evolution (the means by which variation arises) of which relationship is a central component. Inter-relationship, then, ought to be of the first importance to conservationists. The importance to conservation of relationship is evident in any ecology or conservation biology literature where there is much talk of relationship between species: predator and prey / host and parasite / decomposer and detrivore, as well as interspecific competition and mutualism. Attention, too, is given to relationship within species, such as intraspecific competition. Nevertheless, despite this emphasis within conservation upon the importance of relationship, two aspects of relationship are neglected. The first is acknowledgement that all forms of relationship are ultimately relationships between individuals. And the second is acknowledgement of a more fundamental connection between conservation and relationship: the recognition that through relationship we come to know the world. In this sense, it precedes any intellectual stance we might take concerning the environment, any ordering of the natural world, and, as we shall see, any desire to protect it.

Reed (1994) describes the process, whereby, through relationship we come to understand the world. He suggests that we and other animals inhabit a shared world of meaning in which, through the detection of ecological information, we become aware of 'the affordances of the objects, places and events surrounding us'. Affordances are 'ecological values for the observers, they are opportunities for doing something, for obtaining certain resources, or hindrances such as traps and dangers' (p. 112). He further specifies that 'whereas inanimate objects afford actions (to obtain the values or to avoid the dangers), animate objects afford interaction, and socialized objects afford proper (as against improper) action and interaction'. Reed believes that all animate interaction may be:

> based on the perceiving of a shared environment [and that p]erception of the self in the environment is already a social act if that self is a social creature, with an appreciation of the environment to be shared with others – friend and foe, predator and prey, mate and child, and so on. In other words, subjects are not private, but public and shared – even in the kind of wordless sociality that is found in birds and mammals (p. 119).

Interaction, he explains,

> occurs when two or more agents realize affordances in a mutual way, so that each presents acceptable affordances to the other (predator hides from prey and vice versa; potential mates look attractive to each other; etc.) but in socialized interaction the agents' relations are further constrained by the perceived need to present proper affordances to the other (p. 121).

On this understanding, our dissatisfaction with a zoo experience is to be explained, in part, by the fact that we and the zoo animals are (in most cases) simply *not* interacting, since neither we nor the animals viewed are in a position to present acceptable affordances to the other. Reed's analysis would also show that the chimpanzee Moja – though brought up, and habituated with humans – did not commit an 'improper' action by biting her handler. As Hearne (1987) herself decided, Moja inhabited a very different story to a similarly brought up domesticated animal such as a dog. Her action, in fact, was entirely *proper*. According to Reed, it is through our interaction with other animals that we are provided with both social awareness and knowledge. Such socialised awareness and activities, Reed believes, '*precede* the ability to label and refer to socially constructed categories of objects' (p. 123).

Relationships are crucial to our very understanding of the world. By focusing on the abstract concept of 'species', conservationists are in danger – by neglecting the primacy of relationship – of working with an artificial view of the world in which concepts are given precedence over real individuals. Reed's belief in a mutually accessible world is shared by Ingold (1995). Ingold, however, does not make a distinction between interaction and socialised interaction, believing that 'if some relations are social, then all are; and all life would be social life' (p. 31). Although Ingold's position has great appeal, Reed is probably closer to the truth, because there does seem to be a significant difference, say, between our interactions with domesticated and wild animals.

If relationship is crucial to our getting to know the world, it is equally crucial to our capacity for caring. That our capacity for caring is rooted in our care of particular individuals is convincingly and entertainingly argued by De Waal (1996), who demonstrates that ethical behaviour is as much a product of evolution as any other trait. Examples of sympathy, altruism and rudimentary ethical systems are found in many other species, demonstrating that caring is very much a *shared* capacity. De Waal argues that the source of the capacity to care (succorant behaviour) is in the parent-infant relationship and that caring has spanned out from there (p. 43). But it is not only that the individual relationship has an important place in the evolution or history of our capacity to care, it is also an essential element in each of our lives. It is through caring for and about individuals that we come to care more widely: for humanity and for sentient animals in general. It may only be by a process of mental abstraction, that we can extend our caring attitude to concepts, such as species. It may not be possible for us to have a real or direct relationship of caring with concepts such as species; we may only be able to mentally apply an attitude of caring towards them.

This analysis accords with Norton's (1987) criticism of attempts to widen the moral circle from the intrinsic worth of individuals to the intrinsic worth of species. The detailed development of such accounts has, he says, 'proved disappointing' (p. 167); it is 'very difficult to move from concern for individuals to concern for species'. However, his contention that '[c]oncern for individuals flows against the tide of concern for species' (p. 166) is unconvincing, and my reasons for saying so will emerge in Chapter Seven.

The natural world is, of course, only partly animate; the far greater part is inanimate. Yet our appreciation of, and desire to protect the inanimate natural world (if indeed it can sensibly be separated from the animate) is no less grounded in relationship. There is no sudden, detached arising of a desire to protect, say, a species of orchid, but rather, appreciation of the environment evolves through our experiences in, or our engagement with it. Evernden (1978) writes about Cobb's understanding of a child's appreciation of the natural world that she believes is aesthetic. Likening Cobb's notion of aesthetic to that of John Dewey, Evernden writes that:

> [t]he aesthetic experience, in Dewey's terminology, lies in the relationship between the individual and the environment ... Rather than a subject-object relationship in which the observer parades before the supposedly beautiful view, we have instead a process, an interaction between the viewer and the viewed, and it is in that joint association that the

> aesthetic experience lies. Instead of a detachment from the environment, we have a subtle diffusion into it (p. 18).

So although our early engagement with the inanimate natural world is not between individuals, it is certainly an engagement with particular aspects, or places in, the environment. We are, as Evernden recognises, apt to become attached to familiar places:

> there appears to be a human phenomenon, similar in some ways to the experience of territorality, that is described as aesthetic and which is, in effect, a "sense of place", a sense of knowing and of being a part of a particular place (p. 19).

It is through engagement or relationship with the natural world that we come to cherish it. We can engage with other individuals and with places in a way that we cannot engage with abstract concepts such as 'species'. Admittedly, we can care about abstract concepts such as 'justice', but we do so, I suggest, because we can engage with and care about people who are the actual or potential recipients of justice. This equates to the way in which our caring about species is rooted in our caring about actual (or potential) individual animals. To focus attention away from individuals and relationship and onto species is dangerous for conservation since it severs us from the roots of our capacity to care. It is also disingenuous of the conservationist to suggest that individuals are of little relevance to the conservation cause.

The centrality of the individual relationship is eloquently expressed by Ingold (1995) who sees organic life as an active rather than passive process. It is seen as 'the creative unfolding of a field of relations within which beings emerge and take on the particular forms they do, each in relation to the other' (p. 21). He explains that '[l]ife ... is not the revelation of pre-specified forms but the very process wherein forms are generated'. He believes that 'there can be no specification of the form of an organism that is independent of the developmental context within which it comes into being'. Life forms emerge, then, rather than being specified by a genotype. For Ingold, individuals are the very stuff of life: they create themselves.

Given this understanding of life, Ingold sees fit to reject not only the species concept, but also Darwinian theory. But there is no necessity to ditch Darwin in order to hold at least something like this position. Ingold claims that Darwinian theory requires the possibility of a context-independent specification of form. It is, he says, 'this specification – the genotype – that is said to undergo evolution through changes in the frequency of its information-bearing elements' (p. 21). But this is not so. Darwin was working with no understanding of genetics – and genotype – and did not need the concept to assign organisms to species.[6] Furthermore, his theory of evolution requires the context of the natural world both in order to be arrived-at (understood), and in which to take place. Despite the problems I have highlighted regarding the concept 'species', I have no argument with Darwin's description of

---

[6] It is hardly surprising that Darwin put so much emphasis on the species category since he lived at a time when categorisation /classification was pursued with particular zeal.

the way in which biological diversity has arisen, or with the use of the species concept as a way of articulating that variety. There is no need to deny, as Ingold does, that there is a hereditary component to the evolution of diversity. Darwinian theory does not require that there be a pre-determined form (in the genotype), but rather, that the form will emerge *when the genotype is in context.* It has never been supposed that evolution can take place anywhere other than in the real world, so the genotype as denoting form is a mental abstraction only. The only real correlation between genotype and form is in the individual. It is not so much that genotype dictates form. Rather, the transfer of genetic information, through reproduction, is part of the overall process of change in the living world. Biological diversity, ever-present between individuals, is accentuated through reproductive isolation.

Yet it is worth retaining from Ingold's position the idea that individuals have a degree of plasticity and autonomy, that they can 'play their part as producers as well as products of their own evolution' (p. 21). If species are mere by-products of evolution, then individuals are evolution itself. As Ingold puts it:

> ... [e]very being, as it is caught up in the process and carries it forward, arises as an undivided centre of awareness and agency: an enfoldment, as some particular nexus within it, of the generative potential that is life itself (p. 22).

Although firmly within the scientific tradition, Alberch's (1982) view of diversification and adaptation in evolution as independent processes, (with the former inherently preceding the latter) does seem to have, at least the flavour of Ingold's position. In a discussion about the way in which hybridisation can lead to morphological diversification and 'structural disharmonies', Coppinger and Schneider (1995) quote Alberch as stating:

> the role of development in evolutionary processes of morphological change is twofold. On the one hand, the structure of the developmental program defines the realm of possible novelties ... while, on the other, the regulatory interaction occurring during ontogeny can accommodate genetic and environmental perturbations and results in the production of an integrated phenotype (p. 42).

Accommodation, the two authors say, 'is the fitting together of the various organ and behavioural systems'. They give an example from the embryologist Twitty (1966), who 'transferred the eye germ cells from a big salamander species to a tiny species, resulting in a little salamander with great big eyes'. The remarkable thing about this was that 'the little salamanders grew eye sockets that fit their new eyeballs and added extra brain cells to "accommodate" their large eyes'. In a similar way, a study of the evolution of the dog has found that 'the dog's skull has features that have not been selected for but are simply the accommodation of one organ system to another'. Although these examples do not capture Ingold's precise meaning of organism as producer, they offer an alternative expression of the idea, and one that need not reject Darwinian evolution. Individuals not only play a central role in evolution, they have a degree of plasticity, even autonomy. They are not the mere puppets of their genes.

To discover the central role of the individual in evolution, is at once to raise the issue of its worth, as it were, in any talk of the natural world. Traditionally, the worth of the individual has been downgraded, even denied. Value is attached, instead, to species and other forms of 'wholes' such as habitats and ecosystems. I have shown that the concept 'species', is problematic in that it is, just that, a concept. The individual allows of a phenomenological acquaintance with the natural world through the process of relationship, but 'species' offers only an intellectual or acquired understanding of the natural world.

Conclusion

Without the use of the species concept – or something equivalent – we would find it very difficult to comprehend and exchange information about the immense biological diversity of the natural world. The species concept provides a convenient labelling system. But it is vital that those working in conservation recognise the concept as the convenience that it is, rather than as a precise measure, as it were, of the boundaries of the natural world. The species concept is, along with other concepts, an abstraction. Conservation activities, then, ought not to be dictated by any rigid adherence to species boundaries, although it is reasonable that they be guided by them. As a corollary of this, the importance to conservation of the individual and of relationship ought to be recognised and given more prominence. The traditional view of species as the primary natural unit that happens to be made up of individuals is therefore reversed. The concept of 'species' might coherently be seen as something that can map onto the individual, to aid in the transfer of information. As in other contexts, so too in conservation, the individual ought to be regarded as a knowable, concrete unit of importance. Conversely, the concept of 'species' ought to be recognised as a tool for allowing us to access natural diversity in a form that is, if not fully comprehensible, at least generally useful.

With this new understanding of the species concept, what are we to say about conservation measures that are geared towards the prioritisation of species, such as the practice of species reintroduction or specific measures to conserve extant species? With the species concept as a useful guide to conservation policy-making, all things being equal, the practice of species reintroduction is a positive approach to the loss of biological diversity. It has particular merit in that, on concentrating, as it often does, on charismatic species, the human population is usually able to identify well with them. This helps to secure this characterisation, at least, of biological diversity. With the loss of so much biodiversity, the return and safeguarding of any is advantageous and ought to be recommended. Similarly, all things being equal, conservationists' efforts to conserve an existing species – by, for example, providing the necessary habitat or prohibiting shooting – are laudable. These are measures that perpetuate our existing notion of biological diversity.

The problem with species conservation emerges when it is recommended as a conservation measure without regard for anything other than the survival of the chosen species. There are times when the reintroduction or protection of a species would be detrimental to other species or subspecies (and we now know that the former category cannot sensibly be distinguished from the latter). Sometimes, too, individuals are treated entirely without respect, being seen merely as a means to an end rather than, in Kant's terminology, an end in themselves. Contrary to current conservation thinking, individuals are, it turns out, important conservation units. It behoves the conservationist to (at the very least) take their needs into account. The red wolf reintroduction programme, as described earlier, is both harmful to the coyote and arguably disrespectful of the wolves themselves. This, then, is one example of a conservation project that could not be supported in light of this new thinking. Conversely, the sea eagle reintroduction to the Scottish Hebrides has largely avoided such problems and could therefore be sanctioned. If my arguments are accepted, it means that proposed species reintroduction or species protection projects must be thoroughly researched and thought through. It means, too, that the safeguarding of a particular species should not take automatic priority over the safeguarding of other species or subspecies, or the well being of individual animals.

Chapter 7

Wildlife rehabilitation as conservation strategy?

Introduction: the significance of relationship

I have argued that conservation practice ought to value the individual. It should not, as it currently does in almost every case, prioritise without question holistic categories (such as species) over sentient individuals. Such individuals can be coherently defined, recognised and understood. They are meaningful to us in a way that holistic categories – abstractions from the real world – are not. 'Individual' is the kind of category that cannot reasonably be replaced by holistic categories, even if there are good conservation reasons to give holistic categories some prominence.

But it is not merely that the individual is meaningful to us. Because they are foundational to *relationship*, individuals play a vital role in conservation. Ecological relationship, as the source from which biological diversity springs, is the nerve centre of conservation. Since ecological relationship is based upon interaction between individuals, knowledge of and respect for such interaction is crucial to conservation. And equally, since it is through relationship that we come to know about the world – including the natural world – knowledge of individuals and their interactions with others is vital to our understanding of nature. Our sense of caring, too, is rooted to a great extent in our own experiences of care of or for individuals. The emotion of caring drives and empowers the practice of conservation. Conservation, then, ought to acknowledge the importance of individual care. When conservationists urge us to care about (and for) species, the feelings they tap into are based upon (but an abstraction from) our real experiences of caring about (and for) individuals. For the conservationist to remove all focus of attention from the individual is both disingenuous as well as dangerous. It severs us from the very roots of our capacity to care.

But perhaps conservation does not need to concern itself with the notion of care. Perhaps its principles can be adequately justified by, and carried out with reference to, say, utilitarian considerations. One of the arguments offered by conservationists is that we ought to protect species from extinction as they may prove to be a valuable resource in the future. The problem is that utilitarian reasons of this kind can all too easily be usurped by alternative utilitarian reasons, such as: the habitat of this particular species will provide agricultural land for many hungry people. Bryan Norton (1987) demonstrates that utilitarian (or demand) values provide inadequate justification for conservation, arguing that 'demand values provide a shifting and ultimately unreliable basis for a policy of preserving species' (p. 236). Unless our feelings of care have been sufficiently awakened to generate feelings of care for a species, unless we have a desire to protect that aspect of the natural world *for what it means to us* – there may always be reasons against those given for protecting nature.

True, feelings are notoriously fickle. They cannot always be relied upon and may conflict. Indeed, I have already talked about the problem of human fickleness with respect to valuing species differently. Nevertheless, it still remains the case that strong feelings about nature *per se*, a heartfelt conviction about the value of natural phenomena, is arguably the most robust of all possible positions. Such feelings are not open to bargaining or concession in the way that other positions so often are. And this is the case even where there are inconsistencies about the specifics. Conservationists may argue fiercely about which species is to be saved, but agree fully that protecting nature is what matters. If we disregard the role of caring, we fail to acknowledge the most vital motivating force behind conservation action.

The importance of this phenomenon – and acknowledgement of its roots in individual care – has not escaped some environmentalists. Many conservation groups (such as the International Otter Survival Fund, the Iberian Wolf Recovery Centre and the Whale and Dolphin Conservation Society, to name just three) operate 'adoption' schemes, whereby members of the public can sponsor named individual animals. Likewise, conservation groups attract support through advertisements and appeals depicting individual animals. Notable among them is the World Wide Fund for Nature (WWF). Not only is their popular symbol a panda, but their promotional leaflet entitled 'the end is in sight' depicts the faces of different animals from a variety of species, encircled in gun-sights. This gives us the message that these species are doomed. It seems fair to assume that it is the public's sympathetic responses to the fate of these (and other) individual animals that is being sought. The symbol works because it represents a real individual.

Individual caring relationships are extraordinarily powerful. In a discussion about the widespread appeal of the feral pigeon, Ordish and Binder (1967) suggest that part of their attraction is their constancy, their tendency to return to particular feeding places. In this way, 'the regular feeder of pigeons tends to get the same birds every day' (p. 57). This simple feeding relationship is the basis of a strong desire to protect the species, and, as Ordish and Binder disclose, the protectionist lobby is very powerful indeed. For instance, in a press report (France-Soir, 1959) it was revealed that though Paris was to reduce its pigeon population by 300,000, 'the birds were not to be killed; they were going to be caught by firing a harpoon net over them, then fed at a rehabilitation centre [after which] they would be sent to other towns in France' (pp. 74–75). Such an example bears testimony to the passion and power engendered by individual relationship. As the authors maintain, '[t]he emotions are part of the ecological background of man and can have as much effect in establishing the pattern of his life as any physical characteristic' (p. 56). It is important, then, for conservationists to recognise the primacy of relationship. It is important for them to recognise that relationship inspires deep concern and a wish to protect. This is the very backbone of the conservation agenda.

The significance of individual relationship to conservation cannot be overemphasised. Not only is relationship an important root of caring, paying attention to relationship makes good ecological sense. The conservation cause suffers from the problem that it is so vast, so overwhelming an issue to resolve. This leads to inaction, a paralysis born of helplessness. But concentrating on 'the big picture' is sometimes a mistake and a more helpful human perspective may be that of relationship. Appreciating the power of relationship releases us from paralysing inaction. It also serves to re-focus our attention. It reminds us that, after all, the minutiae of life matter, and are worth paying attention to.

The childhood experience of a Scottish musician (Peel, 1995) illustrates this point. Born in 1908 into an illiterate tinker family, he tells of a strict upbringing in which he was forbidden to harm any insects or pick any wild flower. When fishing, he was allowed only to catch sufficient for that meal, and should he arrive home with excess mussels he had to return and release these. Once, after collecting a bucket of water from nearby, his father noticed that it contained a spider. Rather than tip it out, the father ordered the boy to take it back to where he had found it. How, he asked, could he expect there to be clean water for next time if he did not leave the spiders there?

An appreciation of relationship, however apparently insignificant, is important if we want to effect change. It is precisely because we have neglected to consider the importance of each and every relationship with the natural world that the need for conservation has arisen at all. It is no coincidence that native American Indians, or Australian Aboriginals – for whom nature conservation is (generally) not a necessity – consider relationship with the natural world to be of supreme importance. For such hunting peoples:

> ...the details of their beliefs may vary, but [they] share an extraordinary similarity in their attitude to the life around them. Daily experience confirms to them that human life is simply an extension of the natural world and this gives rise to almost universal feelings of equality with it (Dower, 1991, p. 120).

Some hunters identify so closely with other life that 'plants, animals and forces of nature ... are adopted as tribal emblems or totems [where] ... [m]embers of each clan call themselves by an animals' name and believe they share a common ancestry'. This is the ultimate close relationship, one step beyond kinship, in which '[t]he aborigines feel that, as well as being human, they are the totemic animal'. Aboriginal peoples do sometimes cause environmental degradation, but such instances would seem to be the exception rather than the rule. Today's conservationists could do worse than emulate this approach to nature, in which conservation problems are tackled from the bottom up rather than the top down.

The importance that the caring relationship has to conservation can be summarised in the following four points. Firstly, it is through relationship that we come to know about the world (including the natural world). Secondly, the individual caring relationship is an important historical precedent to humanity's sense of caring towards the abstract or the general, including 'species' and 'the environment'. Thirdly, individual relationship is, if not *the*, then *a* very important route by which each of us comes to care more widely. And lastly, paying attention to individual relationship is an important ecological ground rule, the revival of which could significantly advance the conservation agenda.

To summarise, the importance of relationship to the conservation task is evident. It is disingenuous of conservationists not to acknowledge, more explicitly, the importance of the individual. The individual ought to be seen as something more than a representative of a type. It would be foolish to try and disconnect conservation from our natural desire to interact with individuals of other species, as that would sever conservation from the very source of our ethical feelings. Conservation cannot function in an ethical vacuum; it would be more fruitful for conservation policy to reflect the value of individual relationship. This leads to the interesting thought that conservation may have more in common than generally supposed with the kinds of practice normally associated with welfare, that is, practices that prioritise the individual. In order to explore this question, I will consider the case of wildlife rehabilitation, a classic example of a practice that prioritises the individual. Indeed, the welfare of the individual is its *raison d'être*. However, before going on to explore this question, the practice of wildlife rehabilitation requires a brief introduction.

Placing wildlife rehabilitation

Wildlife rehabilitation may be described as the rescue of incapacitated wild animals – that are considered unable to survive in the wild without human intervention – and their subsequent release. However, while this will suffice as a rough description, it is inadequate as a definition, for it leaves a number of unanswered questions. In particular, it is not clear just what should *count* as rehabilitation. Some points of clarification are needed.

To begin with, in order for an action to count as rehabilitation, an individual must receive *direct* help. Rehabilitation consists in specific action for a specific individual. Indirect, or incidental help will not do. In releasing a lactating fox from a trap, it could not be said that her offspring – though clearly saved by this action – have been rehabilitated. In addition, rehabilitation requires that the action performed be considered crucial for the survival or well-being of that individual. It follows from this that it is necessary for a problem to be identified before rehabilitation can take place. Putting food out for the birds over winter as an 'insurance policy' against their death cannot be considered to be rehabilitation. However, should one of these birds be found huddled in a corner one morning, it is a candidate for rehabilitation.

Rehabilitation could be described as being necessary when an animal is disadvantaged. 'Disadvantaged' animals are those that are suspected of being in imminent danger of dying, those whose lives are apparently under threat, or those whose quality of life in the wild is seriously compromised. There are five categories under which prospective rehabilitation may come:

1. Illness, e.g. enteritis;
2. Injury, e.g. a broken limb;
3. Incapacity, e.g. oiling;
4. Dispossession, e.g. disturbance from hibernation;
5. Orphaning (which covers abandonment as well as the actual orphaned state).

Though categories are useful methods of clarification, clearly some cases will straddle more than one category.

A further point relates to human intervention, in particular whether there is a minimum requirement for an action to count as rehabilitation. The degree of human intervention in the process of rehabilitation varies greatly, and can be categorised in the following way:

- De-training. This may be necessary for animals that have become too tame while in captivity and have acquired inappropriate behaviours. It is more commonly required in reintroduction procedures but also with, for example, dolphins rehabilitated from dolphinaria where learned behaviours may need to be extinguished.
- Re-training. After a length of time in captivity, wild behaviours may need to be re-trained so that the animal is capable of survival. Examples of this would be the 'hacking-back' and flying of raptors.
- Rearing, including hand-feeding. This applies to all orphan and some young animals. It usually takes place on rehabilitation premises but may sometimes be undertaken at the animal's place of origin, for example, with fox cubs.
- Treatment, of varying intensity and length. An example of short-term treatment would be the de-oiling of a lightly oiled, healthy bird, while an example of long-term treatment would be the operation, medication, and physiotherapy required for a raptor with a wing fracture.
- Sometimes, all that is required to set a disadvantaged wild animal on its way again is intervention of the most minimal kind. Grounded swifts are not infrequently brought to rehabilitation centres. Often, if the bird is held aloft on an open hand, it will take off, an action the swift is unable to do from the ground. The question is should this be classed as rehabilitation?

In her report entitled *A Review of Live Strandings of Cetaceans: Implications for their veterinary care, rescue and rehabilitation in the UK,* Mayer (1996) implies that the refloating of stranded animals is not rehabilitation, for these actions are categorised separately. It is not clear why this is so. I can only suppose that either the intervention is of too minimal a kind, or of the wrong kind, or that it is discounted as rehabilitation because it does not take place within a rehabilitator's premises. Should actions such as these and the relaunching of swifts be counted as rehabilitation?

The modes of human intervention that are universally accepted as constituting rehabilitation are training, rearing and treatment. What is common to all is some kind of *correction procedure*, a procedure that changes the expected course of the animal's life, thereby allowing it to resume its earlier mode of being (wild survival). There seems no reason, then, to discount relaunching as rehabilitation on the basis of its particular kind of intervention: it is every bit as much a correction procedure as the other types of intervention.

To recap, wildlife rehabilitation can be so-called when: a free-living wild animal's survival or well-being is recognised as being seriously threatened; and when human intervention to remedy this involves a correction procedure of some kind, aimed specifically at the animal in question, and enabling it to return to its wild existence.

It could be argued that rehabilitation practice is fast becoming a science. There are many within rehabilitation – particularly those in the veterinary profession – who are passionately endeavouring to standardise, rationalise, quantify: to create set formulae for all occasions. To a degree, this has to be applauded: for too long rehabilitation practice has lacked any standardisation. Certainly, there are aspects of rehabilitation that are amenable to scientific rationalisation, in particular, the veterinary input. But other aspects of rehabilitation are more akin to an *art*. They involve that sort of expertise that requires 'know-how' and intuition. No amount of theorising or use of formulae will enable a rehabilitator to assess, for example, when an oiled bird is ready to be cleaned, when an orphan can safely be left to eat on its own, whether a particular individual is coping with captivity, or when an individual is ready to be released. The RSPCA's recent idea of developing a point's system to assess whether an individual should be rehabilitated or not (Powell, personal communication), while a reasonable attempt to introduce rigour, is therefore sure to fail. As with much other work involving animals, rehabilitation requires sensitivity and an open mind; it is not entirely amenable to rigid rules. What rehabilitation practice needs are staff members who have a *feel* for the animals, as well as knowledge and expertise.

Wildlife rehabilitation: current attitudes

The attitudes of rehabilitators diverge in many areas, not least concerning the wider impact of rehabilitation. In particular, there is a disagreement about its value to conservation. It is probably true to say that the majority view amongst rehabilitators is that rehabilitation is of little conservation value, being seen primarily as having an animal welfare function. It is often acknowledged, though, that there are incidental or indirect benefits to conservation. The way in which rehabilitation draws attention to environmental degradation – say, oil pollution – would be one example. Very occasionally, rehabilitation is recognised by rehabilitators as having direct conservation value (the Irish Seal Sanctuary, based in Dublin, believes the two practices work hand in hand) and more will be said of this in Chapter Eight.

The public's attitude towards rehabilitation is mixed, but seems to be largely positive. Where criticisms are made they are generally concerning the use of funds. It is sometimes argued that the money spent on rehabilitation could be better spent in being directed towards the protection of wildlife in general. Critics of the project to rehabilitate Keiko – the killer whale star of the film 'Free Willie' – said that money would be better spent on wider conservation projects (Phillips, 1996). This is sometimes a holistic argument (species or habitats are more important than individuals) but frequently, it reduces to a utilitarian argument: it is a question of numbers. However, the number of rehabilitated animals is substantial. Kirkwood and Sainsbury (1996) report that:

> In Britain alone, there are several hundred individuals and organizations involved in treatment and rehabilitation of wildlife casualties ... and a recent survey suggested that at least 16,000 wildlife casualties were taken into captivity for treatment annually in Britain ...

They suggest that this figure 'is probably a considerable underestimate' and go on to illustrate the scale of rehabilitation elsewhere. In the Netherlands, wildlife hospitals are estimated to deal with over 30,000 birds a year, a high proportion of these being released. Kirkwood and Sainsbury believe that '[a] significant proportion of the individuals in populations of some species are treated in wildlife hospitals at some stage of their lives' (p. 237).

Though negative public responses are encountered, positive responses are more common. Rehabilitation centres are largely funded by public donations, and pleas for donations for specific purposes are often responded to with enthusiasm. The public is also very willing to become involved in the rescue of wild animals; people will often make substantial efforts to save them. One man who found an injured mouse travelled several miles to the RSPCA centre at West Hatch in Taunton in the hope of its being saved (Powell, personal communication).

Scientists and conservationists respond to rehabilitation in a range of ways. It is not uncommon to find conservation groups working alongside wildlife rehabilitators. The RSPB, for instance, has worked closely with the RSPCA after oil spills such as at Milford Haven, Wales, in 1996. It is even possible to find one or two cases of conservation organisations actively pursuing wildlife rehabilitation themselves. Just one example is the International Otter Fund. This is a professional, science-based organisation based on the Isle of Skye, actively working to conserve thirteen otter species in the wild. They simultaneously run an otter hospital and sanctuary on the island. There is, however, little academic recognition of the value of wildlife rehabilitation. Probably for this reason, any criticism from academics is taken very seriously indeed. Sharp's (1996) paper criticising the rehabilitation of wildlife as a waste of time and resources, generated great concern throughout the rehabilitation circle. Ian Robinson, veterinary director of the RSPCA's King's Lynn centre, spoke at the 1996 WildAid Conference in Inverness of the importance of addressing criticisms which could be damaging to the acceptance of rehabilitation practice. And James Kirkwood, a veterinary surgeon involved in wildlife rehabilitation, candidly exposes the problems that rehabilitation practice poses both for conservation and for welfare (Kirkwood and Sainsbury). One of the reasons for the lack of academic support for rehabilitation appears to be because conservation is viewed as a practice that ought not to concern itself with individual welfare. Yet it has become increasingly clear that conservation has more reason than supposed to endorse the value of the individual. With this in mind, I will now consider the ways in which wildlife rehabilitation, as perhaps a paradigm case of concern for the individual, can be seen to have conservation value.

Wildlife rehabilitation and conservation: shared ground

I have already established that care – as expressed through relationship – underpins conservation philosophy and is an important catalyst for change. Conservation is not a detached, purely science-driven exercise, it depends upon a deep sense of caring. (Not, of course, that science itself is detached, for as Linzey (1996) points out, it 'reflects the cultural assumptions and values of the age'.) Seen in this light, conservation has a great deal more in common with wildlife rehabilitation than might at first be supposed.

Through the care of individual animals, wildlife rehabilitation offers all-important relationship, a window through which we access the cherishing of species and the wider environment. Dave Phillips (1996), in charge of the rehabilitation project to return Keiko (the killer whale) to the wild, argues that Keiko serves as an icon. Conservationists have little success in raising public concern about the threats to the world's oceans, for, as Phillips notes, it is difficult to capture the importance of such things as phytoplankton. However, the public identify strongly with Keiko, and this special relationship enables connections to be made with the state of the world's oceans. Keiko, as Phillips describes it, 'puts a face on the ocean'.

Wildlife rehabilitation is important as an entree to conservation. But it offers, too, an outlet for the caring or nurturing that (many argue) is important for our well-being. The readiness with which people sacrifice their time, energy and money in the rescue of oiled birds, for example, is not *merely* an indication of selflessness. Rescue, treatment and rehabilitation have a significant 'feelgood factor' too. That this should be dismissed in derogatory terms as mere sentimentality – as Leahy (1996) does – is nothing short of ignorance. Whether one feels compassion for and thus a desire to help other creatures found to be suffering, is, as Tony Banks (1996) so concisely put it, all to do with whether one is a decent human being or not. Conservation does well to align itself with a practice the undertaking of which is associated with human decency. But it is not merely that conservation benefits from the association; it is, in its own right, an ethical practice. Conservation practice seeks to perform right action. Wildlife rehabilitation and conservation are, then, like-minded practices.

From theory to practice

So far, I have sought to establish the way in which relationship – which characterises wildlife rehabilitation – is the foundational linchpin of conservation. In effect, what I have established is the *theoretical* link between the two. This is not the only kind of link to be found. Where relationship is honoured through wildlife rehabilitation, there are a number of *practical* ways in which conservation benefits. By way of an introduction to these links, I will consider the degree to which the defining lines between wildlife rehabilitation and certain conservation strategies seem to blur.

Links with restocking and reintroduction Conservationists have not, traditionally, practised wildlife rehabilitation. They have, however, been extensively involved in two other practices – reintroduction and restocking – as well as, to a lesser extent (though seldom recently) introduction. The distinction between the terms has not always been clear, but in the *Manifesto on Animal Reintroductions* (Boitani, 1976), the authors offer definitions for each. Introduction, they describe as the '[r]elease of animals of a species into an area in which it has not occurred'. Priority should, they claim, always be given to reintroduction, defined as the '[r]elease of animals of a species into an area in which it was indigenous until exterminated as a consequence of human activities'. The third area of involvement – restocking – a practice undertaken to help conserve dangerously reduced populations, is defined as the '[r]elease of animals of a species into an area in which it is already present' (pp. 229–300).

In light of these descriptions of legitimate conservation practices, it is less obvious that rehabilitation is as much of an outsider as is frequently supposed. Where rehabilitated animals are returned to the area from which they originated, rehabilitation is a means of helping to maintain the population. In cases where restocking might be considered, rehabilitation offers a less manipulative option that will complement restocking programmes or perhaps even pre-empt them. Where, on the other hand, rehabilitated animals are released in an alternative suitable area to the one from which they came (as is sometimes unavoidable), rehabilitation seems little different – albeit on a smaller scale – to either restocking or reintroduction. In the case of young animals, it is often impossible to release them in their area of origin (perhaps because this information is not known) or it may be inappropriate (they may, for instance, be attacked by resident animals).

The release of young animals very often, then, falls into the categories of restocking or reintroduction. A suitable site for young foxes, for example, would be one that can sustain the species and so the likelihood will be that there are already foxes around. (If no foxes are to be found, the chances are the site is unsuitable). The rehabilitator's intention would be that the young foxes, though needing to seek out their own territory, integrate with the existing population. This does not seem appreciably different from the practice of restocking.

Taking another example – that of badgers – the situation is a little different. While rehabilitated badgers may be returned to the population from which they came, they are not generally released into a different population. The worry is that resident badgers may harm or even kill intruders.[1] For those badgers that cannot be returned to their place of origin, rehabilitators need to find suitable sites from which badgers have disappeared or in which badgers have always been scarce. The south-west of England has a very large badger population and so rehabilitation centres receive many orphans. The south-east, on the other hand, while offering suitable badger-habitat, has a very small population. (This is thought to be the result of very heavy keepering – and thus killing of badgers – pre-war.) In the past, badger cubs from the south-west were not infrequently rehabilitated into the south-east. (Unfortunately, due to the belief that badgers spread TB in cattle, rehabilitated badgers can now only be released into restricted areas.) As the situation was, though, it once again seems remarkably similar to the conservation practice of restocking. The example also serves to illustrate the difficulty of distinguishing sharply between restocking and reintroduction. It is not clear how large an area must be for the release of animals to count as reintroduction as opposed to restocking. While every group of badgers must be released into a territory that is free of other

---

1 Due to the difficulty in finding suitable release sites, the RSPCA is proposing to release a small number of badgers at a time into areas already populated by badgers (Powell, personal communication). This is controversial and is not viewed favourably by all. Some feel that further field studies should be undertaken in order to assess the likelihood of success.

badgers, in some cases there will be badgers in the next territory (arguably a case of restocking), while in other cases there may not be other badgers nearby. It could be argued that this latter situation is a case of reintroduction. The query arises because of the vague nature of the term 'area' used in the definitions. So it may be that in some of these examples, rehabilitators are practising not just restocking but reintroduction, which, it is worth noting, is the conservationists' *piece de resistance.*

Other examples of rehabilitation fall more clearly still into the category of reintroduction. Whitcher (1995) noted that there were plans afoot to release badgers onto the island of Skye, north-west Scotland. Since the badger became extinct here in the 1970s such a project could not fail to be considered a reintroduction. Consider, too, the case of rehabilitated orang utans. Because of worries over disease spread and overpopulating particular areas of forest 'Herman Rijksen from the Institute of Forestry and Nature Research in Wageningen, Holland is now trying a different approach in Kalimantan' (Western, 1994). Confiscated orang utans, after six months in quarantine are:

> Held in groups of 10 to 15 animals so that they can develop a social structure, and then taken as a group to a suitable area of forest where there are no longer wild populations. Here they are released.

It could be argued, then, that some cases of reintroduction, at least, are not easily distinguishable from recognised conservation strategies. Yet there is a stronger point, still, to be made. Conservationists are keen that 'conservation of existing populations and ecosystems should always take priority' over other forms of conservation (Boitani, p. 300). This being so, in its role of helping to maintain existing populations rehabilitation may even be argued to *better*-fulfil conservation objectives than some of the conservation practices currently employed. Such practices include some of the reintroduction programmes that require a high degree of interference, manipulation and control (see Chapter Nine). This is particularly relevant given the recent 'hands-off' approach among some rehabilitators in dealing with selected orphans, such as fox cubs. Routh (1995) formerly of the RSPCA Wildlife Hospital in Cheshire advocated caring for orphaned fox cubs *in situ* rather than taking them to centres to rear before release.

Rehabilitation must also be seen in a positive conservation light given that it makes no drain on existing populations in the way that reintroduction and restocking inevitably do. Seen from a conservation perspective, rehabilitation might be said to be making use of what would otherwise be 'lost stock'. Indeed, there are two important ways in which rehabilitated animals may serve conservation by providing such 'stock'. The first is where the species is exceedingly rare. Rehabilitated animals can therefore account for a significant proportion of the entire population. For example, there is a proposal to use rescued and rehabilitated monk seals for a captive breeding project of this endangered species. It is felt that captive breeding 'should not add to the threats facing the small populations that survive' (Lavigne, 1994) and such a scheme provides a way of honouring that sentiment.

The second way in which rehabilitation can provide 'stock' for conservation projects is where local individuals are required. The barn own is one such example. The 'Department of Environment Code of Practice for the Release of Captive Bred

Barn Owls in Britain' stipulates that one of the criteria for reintroduction is that captive stock should be '[f]irst generation progeny of disabled wild birds found within 200 km of the proposed release site, wherever possible'. In other words, such local, disabled birds represent the best possible choice for captive breeding. The captive breeding and reintroduction of barn owls is not, however, a straightforward success story. Barn owls are very easy to breed in captivity, and consequently cheap and readily available. Breeding and release programmes began to appear in the early 1970s in all corners of the country. Release programmes increased until '[i]n 1992 it was estimated that between 2,000 and 3,000 birds were being released annually by about six hundred operators' (p. 2). Many of those involved were inexperienced and their methods shoddy. Barn owls were being released only to die, and there was little real concern for conservation. It was against this backdrop that the Code of Practice was formulated by the Department of the Environment. The situation has since improved.

It is worth noting that if the reintroduction of barn owls very often has not been successful, it is because the programmes were poorly conducted, not because the idea of barn owl reintroduction is fundamentally flawed. In the Barn Owl Trust's *Barn Owl Re-Introduction Report 1986–1988*, David and Frances Ramsden conclude, after a thorough enquiry, that barn owl reintroduction, if done properly, *can* work. They say:

> Habitat is the key to long-term Barn Owl conservation; however, conservation is not at present given sufficient consideration in land management. Should the day ever arrive when Barn Owls no longer need active encouragement, so much the better. Until then, habitat improvements, provision of nest-boxes, protection of breeding and roosting sites and reintroduction (where appropriate) can all make a contribution towards 'saving the Barn Owl'.

Where barn owl reintroduction is considered appropriate, birds taken in for rehabilitation but unable to be released are the best candidates for use in the captive breeding programmes. In this way, rehabilitation and reintroduction work very well together. In cases where rehabilitated barn owls can be released, rehabilitation must be seen as an increasingly important conservation tool in its own right. Released individuals help to maintain existing wild populations.

Although not traditionally placed in a conservation context, rehabilitation does, in fact, share much common ground with the practices of restocking and reintroduction. In some cases, at least, rehabilitation is indistinguishable from either of these two practices. To many, this will be a surprising finding because rehabilitation is regarded as having animal welfare – but not conservation – value. Can this conservation role that I am suggesting for rehabilitation be supported in other ways?

Shared methodology Not all cases of rehabilitation will have direct conservation benefits. Only some rehabilitation will provide 'stock' for captive breeding schemes, and a proportion, only, of rehabilitation can be easily described as restocking or reintroduction. For example, much rehabilitation involves animals of species that are not endangered and therefore would not be candidates for restocking or reintroduction. And often, the numbers of animals released at any one time will be smaller than the numbers generally released in restocking or reintroduction programmes. Yet even in these cases rehabilitation benefits conservation, and that is through the sharing of knowledge. The methods and techniques used in rehabilitation are very similar to those used in reintroduction. Of particular relevance will be knowledge relating to release techniques and the preparation of animals for coping back in the wild. As the late Gerald Durrell's organisation declares in a Jersey Wildlife Preservation Trust promotional leaflet, '[r]eleasing captive reared animals into the wild is a new skill for which there are few general rules'. There is a great need for further knowledge, and this, quite obviously, is transferable both ways. One example of shared methodology is revealed in the summary of a discussion reported in *Reintroductions: Techniques and Ethics*. The author believes that:

> Probably the best technique to reintroduce successfully the woodpecker is to raise young birds in an aviary situated in a suitable place for their release, which occurs later, after training them to use special feeding mangers until they become gradually independent (Boitani, 1976, p. 236).

This is precisely the method used in rehabilitation to return various species of animal and bird back to the wild. It is variously termed 'soft-release' (as opposed to 'hard-release' where the animal is set free with no back-up feeding or care), and 'hacking-back', a term used mainly for birds of prey. Rehabilitation has been, and will continue to be, of benefit to conservation, in that skills gained here are transferable to reintroduction projects.

The rehabilitation of common species is clearly important in its own right, yet the knowledge gained from this work can substantially benefit endangered species. Rehabilitation is a relatively new practise and the methodology is still subject to a degree of trial and error. There are inevitably casualties along the way. If rehabilitation were only practised with members of endangered species and not with common species, fewer endangered animals would be saved because the entire learning process would have to take place on these individuals. As it is, progress made towards successful rehabilitation methods with individuals from common species, paves the way for greater success in general. This includes successes with individuals from endangered species. Routh (1996) believes this to be true of rehabilitator's work with bats. He writes:

> The care and rehabilitation of bats has made great advances in recent years ... the formation of the Bat Conservation Trust and more recently the publication of *Bat Care News* means that discussion and interchange of ideas is occurring with greater frequency, all of great benefit to bats. Skills learnt on other types of wildlife can be used to good effect on bats and the skills learnt on the common species will always stand one in good stead should a rarer species appear. Perhaps if there had been skilled rehabilitators around at the time then some species now confined to the history books would still be with us today.

Likewise, the success rehabilitators have had in cleaning oiled Jackass penguins (Maynard, 1994) (an endangered species found off the Southern African coast) must have been due, at least in part, to the knowledge acquired by rehabilitators regularly removing oil from common species. The ringed seal (very rare in British waters) that was treated at the Orkney Seal Rescue in 1993 (Anon, 1994) owed its life, almost certainly, to the existence of the rehabilitation centre and the staff's expertise in dealing with sick common and grey seals. The ringed seal is almost impossible to distinguish from the common seal, except on close inspection. It may not have been noticed let alone identified, but for the existence of the rehabilitation centre and its trained staff.

It is important, then, that rehabilitation continues for all species if endangered species are to receive maximum benefit, because the learning process is ongoing. It is sometimes the case that conservation gains substantially from the rehabilitation of endangered species. Rehabilitation of the endangered monk seal provides a good example. On the island of Alonissos, a rehabilitation centre has been set up jointly by the Hellenic Society for the Study and Protection of the Monk Seal (HSSPMS) and the Seal Rehabilitation and Research Centre (SRRC) in Pieterburen, Holland (Johnson, 1993). Abandoned pups are rescued, treated, and, when old enough, released to help boost the small populations. It was reported, in 1993, that plans were afoot for a new and larger rescue centre. Dr. Peter Reifinders, chairman of the IUCN seal specialist group, commented that the ten to twenty orphan seal pups that could be saved annually by the establishment of a new rehabilitation centre 'would be a very significant number, given that the population is so low'. Johnson suggests that these rehabilitation efforts could be the saving of the species. Certainly, this would seem to be a better, less intrusive way of helping the species than a captive breeding programme.

For many endangered species, though, loss of habitat is the greatest threat to them. In such circumstances, it may seem that rehabilitation could have nothing to offer. But this is not the case. Sometimes, rehabilitation will provide the incentive for protecting all-important habitat. Borner's (1979) work in rehabilitating orang utans (a species threatened primarily by loss of habitat) was, she felt, worthwhile from a conservation point of view. In an appraisal of the wider effects of her efforts over the years, she writes:

> I was encouraged to think that the idea of conservation has come to stay in Indonesia. The current 5-year plan, for example, provides for doubling the total area of reserves and – with international support – for improvement in the condition and standards of monitoring and protection in existing parks (p. 175).

Similarly, when Dr. Birute Galdikas was asked what she felt had been her achievement in her work with orang utans in Borneo, she replied:

> From a scientific point of view, I've been following the lives of wild orang utans, but what ultimately matters is that just by working here and rehabilitating orang utans, a bit of tropical rainforest that would have been destroyed has been saved (Cater, 1990).

Education and raised profile There are also more subtle practical conservation benefits to be gained from rehabilitation. These are the benefits gained from raising the profile of conservation issues. Rehabilitation centres have direct contact with the public. Whether or not they take an active part in education, they cannot help but raise the profile of environmental issues. For example, by simply existing, they alert the public to the presence of local wildlife. They also raise awareness about the fact that wild animals often require rescuing, thereby highlighting environmental problems such as oil pollution. The rehabilitation of the killer whale Keiko – through appeal to feelings of empathy, compassion and care – has had the important side effect of alerting people to the oceans' many threats. It is not just that individuals are a route through which to teach us about conservation; they are the most appropriate route. This is because we are creatures that have evolved to pay attention to other individuals; we are social animals. Rehabilitation centres and individual projects help to raise the value of local wildlife by promoting the idea that such rescue is worthwhile. Though people are surrounded by wildlife they may nevertheless be unaware of it and remain unfamiliar even with common species. For instance, a reasonably frequent occurrence is when members of the public mistake rescued squabs (young pigeons) for birds of prey (young pigeons do look rather vulture-like!). By simply existing, rehabilitation centres engender a positive approach. Should an injured animal be found, there is somewhere to take it where it will have a chance of recovery. At a time when there is much that is negative, even gloomy, in environmental terms, it is important that there are avenues of hope.

Rehabilitation centres have the opportunity, then, to actively educate the public, not only about animal welfare but also about natural history and environmental issues. A number have educational material in the form of display boards, videos and leaflets. Others offer talks and guided tours. The rehabilitation centre at Lower Moss Wood in Cheshire funds itself by providing educational visits to (mainly) schoolchildren at its seventeen-acre woodland site that has nature reserve status. Here, efforts are made to focus children's attention on the importance of protecting local habitats. Similarly, at Mallydams Wood, in south-east England, the RSPCA combine nature reserve (offering residential ecology-based courses) with a wildlife treatment facility.

Some conservation organisations have recognised the educational value of rehabilitation. The 'Sea Turtle Protection Society of Greece' launched a 'Rescue and Rehabilitation Scheme' in 1991 which sought to involve local people in rescuing stranded turtles. Through active involvement in treating injured, stranded turtles, the local people have become sensitive to their plight. This has had a knock-on effect as regards the turtles' general protection, for fishermen (previously one of the turtles' greatest enemies) are now more sympathetic towards them (Nantsou & Antipas).

Borner (1979), in her work in Sumatra rehabilitating displaced orang utans, believes that her efforts amounted to more than the rescuing of numerous individuals. By providing a centre to which the local authorities could now send confiscated orang utans, enforcement of the law prohibiting the keeping in captivity or tracking of these creatures could, for the first time, be enforced. After habitat destruction, trading was the orang utans greatest threat. Borner writes that '[t]he results were indisputable. Trading in orangs had virtually come to an end in northern Sumatra and on the west coast of Aceh ...'. The educational value of the project is clear from Borner's observation that '[b]y now most of the inhabitants of both provinces had heard of our station, and knew that the law concerning orangs was being taken seriously' (pp. 174–5). If such a project is valuable in raising the consciousness of local people and governments, its influence may not end here. Borner muses that:

> Perhaps, too, our presence had helped to convince the Indonesian Government that the world outside Indonesia regarded it as custodian of the orangs, and was interested in their welfare. So the Government might be persuaded to use the legislation at its disposal, to strengthen the agency (the PPA) directly responsible, and even to ensure that its timber resources were exploited in ways which were not entirely destructive to the orang populations (p. 175).

The orang utan rehabilitation work of Dr. Birute Galdikas in Borneo is closely linked with her scientific study of wild orang utans. Recognised as an international conservationist and expert on the wild orang utan, she has also, for many years, been involved with their rehabilitation. Her organisation, The Orangutan Foundation, offers the public the opportunity for participation in research at the camps in Borneo; 50 percent of the cost goes towards the orang utan work. There can be little doubt that the chance of interacting with orang utans in the process of being rehabilitated is a significant draw. The conservation work benefits both financially and practically from public participation, and the conservation message is carried to many different parts of the world.

Work in Zambia with chimpanzees has similarly beneficial educational spin-offs. Though the chimps are not released into the wild, they live – at the Chimfunshi Wildlife Orphanage – in large, naturalistic enclosures. The existence of this orphanage 'has encouraged the Zambian government to confiscate chimpanzees entering the country illegally from Zaire' (and when sold at local markets or held illegally). Such rescue centres can 'be powerful educational programmes' (McGreal, 1993).

Another way in which rehabilitation raises the profile of conservation issues is in introducing the public to individual animals that, for whatever reason, have been unable to be released back into the wild. Amongst rehabilitators, this area remains controversial. But there is no doubt that where it does occur, there can be conservation benefits. It can be particularly useful to conservation when the species in question is widely misunderstood, as is the case with the various bat species. Since bats are declining as a result, largely, of loss of roost space and use of poisonous timber treatments, it is vitally important for their survival that the public become aware of these issues. Bats increasingly rely on buildings in which to roost, as their natural roosts (such as old trees) decline. But the plight of the bat will fall on deaf ears if it is viewed – as many still think it to be – as a pest, a source of disease, or just plain repulsive! Corporate funding, which now helps support the red squirrel, has not been forthcoming for any bat species. By introducing householders to a tame bat – especially one that is unable to fly – local bat groups have found that fears and misapprehensions can be sufficiently allayed. This can, for instance, lead to residents allowing bat roosts to remain in their houses rather than demanding their exclusion.

Where rehabilitation takes place, there is a concurrent raising of environmental consciousness, if not direct environmental education. Increased awareness may touch local people, local governments, and even the international community. The practical outcome may be profound. In general, though, the educational conservation benefits that flow from wildlife rehabilitation are small-scale and local. These are nevertheless of great importance and represent the tackling of conservation problems from the bottom up.

Individual care fulfils conservation objectives

I have identified various practical conservation benefits that accrue from the practice of wildlife rehabilitation. But one final benefit remains to be identified. I earlier argued that the individual is important to conservation in being the primary source of – and for – caring. Caring is crucial for galvanising us into action to protect the environment. But caring for individual animals yields an additional, more practical value to conservation. Paying attention to, and getting to know individuals makes good conservation sense. Indeed, through individual care we are fulfilling, precisely, the aims and objectives conservationists promote.

This applies to scientists working to distinguish very similar species – that is, in cataloguing biodiversity. It also applies to conservation methodology. At a Wildlife Symposium in Cheshire, Tony Warburton (1995) made the point that we need to get to know the individual animals (and learn about their inter-relationships) in order to understand the species. He gave the example of owl calls. Individual owl calls differ, a fact often overlooked, but if one were to take the time to distinguish between the calls, much could be leaned about the behaviour of the owls. Similarly, Ginsburg and Cole (1994) write that '[t]o understand the population dynamics of wild dogs, a scientist must be able to identify individuals and follow them through their lives'. Jane Goodall's study of the Gombe chimps is another case in point. She learned a great deal more about chimps through getting to know the individuals than was previously known about them. It is this kind of meticulous study that distinguishes a good naturalist from a mediocre one. And since natural history informs conservation – often providing the sort of information without which conserving a particular species would fail – it makes good conservation sense to pay particular attention to the individual.

Wildlife rehabilitation's concentration upon the individual has benefited conservation in this way. Many years of orang utan rehabilitation work in which known individuals are followed and studied, has left us with a greatly increased knowledge of their behaviour and needs. Monica Borner (1979) describes how she watched a mother (an orang utan rehabilitated back to the wild) interacting with her new infant:

> Binjai seemed to be a relaxed, happy mother; used to the company of humans, she did not in the least mind being watched and followed through the forest. So I visited Ketambe a

> month after the birth, and again every 5 or 6 weeks for the first 5 months, to see what she did with the baby, and monitor its development (p. 93).

Borner gleaned new information relevant to rehabilitation. Her study was also groundbreaking in terms of natural history, for normally, wild mothers with young are impossible to watch.

A similar kind of conservation benefit accrued from an American study to ascertain whether orphaned, infant insectivorous bats could be rehabilitated. It has revealed new and surprising information about bats (Fry, 1996). The results of this study suggest that, contrary to the previously held expert belief, orphaned bats may be able to be rehabilitated and returned to the wild. It was not previously thought that such bats could learn to feed or find a roost without parental help. However, Fry found that human-reared infants gained sufficient strength to maintain prolonged flight, were capable of self-feeding, recognised humans as natural predators and were able to locate and use various shelters within their aviary. This is of great interest to rehabilitators But so too does it provide new information relevant to natural historians that may be germane to bat conservation.

In Britain, at least three post-release studies of rehabilitated hedgehogs – in which treated and released individuals were tracked with the use of telemetry equipment – have been conducted. The studies have sought to discover whether, and how well, such hedgehogs survive, and consequently how successful rehabilitation is. They have also provided data about hedgehog behaviour. For instance, Morris *et al* (1993) found that, though 'in warm weather hedgehogs often do not bother to build a proper nest [and thus] do not need to return to the same place each day, or to a previously used place' the study animals usually did so (p. 57). Data collected relates not only to nest-building activity, but also to dispersal and movements, body-weights, courtship, predation, interactive behaviour and diet. It could be argued that rehabilitated animals behave differently from those that have always been wild and that such studies tell us nothing about wild behaviour. Even if it were so that rehabilitated animals sometimes behave differently, this could be just as helpful to those seeking to understand normal wild behaviour.

The potential for co-operation between wildlife rehabilitation and conservation is clear. Consider, for instance, the following discussion about the difficulties of saving the endangered African wild dog:

> Studying highly mobile endangered species such as the African wild dog can present a dilemma as any intervention and handling of wild animals carries risks. But data critical to conservation cannot be collected without intervention – disease screening, genetic studies and fitting radio transmitters to study ecology and mortality of animals all involve giving anaesthetics and/or collection of tissue samples. While no instances have been recorded of wild dogs dying while under anaesthesia, for many species, such as rhinos, just putting an animal to sleep could kill it ... how do researchers balance these risks against the urgent need to collect information critical to conserving the species? (Ginsburg and Cole, 1994).

Though it seems that the handling of wild dogs and fitting of radio collars are not directly implicated in this species' population crash in the Serengeti, the authors concede the possibility of an adverse reaction resulting from 'a complex interaction of factors such as nutrition, stress and the prevalence and virulence of disease not found in other ecosystems'. If such research were to draw upon data taken from animals being rehabilitated (and thus already captive) conservation could benefit enormously, without the risk to the free-living population.

Getting to know the individual animal, then, can have a positive impact on conservation. Knowledge of individuals is often helpful in increasing our ecological understanding; indeed it is often essential if we are to come to know the intricate social lives of some animals. Rehabilitation practice provides an ideal opportunity for the detailed study of individuals, whether this involves observation or more intrusive data collection. Data acquired is often of relevance to natural historians, conservationists and rehabilitators alike. The potential for detailed study of animals, using rehabilitated individuals, is, as yet, largely untapped. Collaboration between rehabilitators and conservation biologists could usefully be developed and extended.

Conclusion

Individual relationship ought to be given greater attention by conservationists. The caring that is so vital to the implementation of conservation activities has an important historical as well as experiential root in individual caring relationships. So there are good reasons to argue that conservation ought to acknowledge the value of those practices dedicated to individual care. Conservationists are disingenuous in failing to acknowledge the theoretical links between the caring relationship and conservation practice. Conservation has much in common with – and in a practical sense much to gain from – at least one such individualistic practice: wildlife rehabilitation. Taking together the difficulties (identified in Chapter Six) connected with prioritising 'species' over 'individual', and the undeniable importance of individual relationship, it becomes clear that there is a need for conservationists to reassess their current outlook. In Chapter Six I argued that holistic categories such as 'species' should have only descriptive and not prescriptive value. We can now add that the individual should be viewed as having value and relevance to conservation in a way that it has, until now, been denied. There should be a shift in conservation thinking. What is needed is a new emphasis that acknowledges the importance of the individual.

Chapter 8

Conservation and individual worth; compatibility: problems and answers

Introduction

In Chapter Seven, I suggested that wildlife rehabilitation ought to be considered a legitimate conservation strategy. It is not only of relevance to conservation, it actively supports the conservation rationale. I argued that conservationists ought to adopt a new approach, allowing for a greater emphasis upon the value of the individual. The common supposition that individual animals are of little relevance to conservation, that conservation is concerned exclusively with groups of things or 'wholes' such as species, habitats and the like, ought to be abandoned.

But if conservation is to take the value of the individual seriously, two worries arise. Firstly, some say that the practice of wildlife rehabilitation, in prioritising the individual (and in particular the *disadvantaged* individual) compromises two values held dear by conservation, namely natural selection and evolutionary fitness. And secondly, there may seem to be an underlying discord between the holistic values of conservation and the individualist values of wildlife rehabilitation. This is a discord that is embodied in the widespread conservation practice of culling, where individuals are bestowed little – if any – value. If my claim that conservation ought to make room for the importance of the individual is to be taken seriously, these worries must be addressed. I will start with the question of whether wildlife rehabilitation, in its capacity of championing the individual, compromises the conservation values of natural selection or evolutionary fitness.

Does wildlife rehabilitation undermine evolutionary fitness?

Conservationists value natural processes in which natural selection plays a pivotal role. Natural selection, it is believed, ensures the 'survival of the fittest', and this, some critics say, is seriously undermined by the sort of human intervention typical of wildlife rehabilitation. Kirkwood (1993, p. 237), for instance, has written that '[t]reating compromised individuals and therefore giving the less fit a second chance represents an interference with natural selection'. Elsewhere, Kirkwood and Sainsbury (1996, p. 239) expand on this:

> Wild animals have always fought their own battles with competitors, parasites, infections, and with the rigours of the environment and are as they are – anatomically, physiologically, behaviourally and immunologically – entirely because of this. The treatment of compromised individuals and thus giving the evolutionarily "less fit" (which by definition is what wild animals that are sick or injured through natural causes are) a second chance, is no less an interference than shooting the fit.

There are serious problems with this criticism. To begin with, the authors claim that wild animals that are sick or injured through natural causes are, by definition, evolutionarily less fit. Here, however, they conflate evolutionary fitness (a hereditary concept) with the popular notion of fitness as 'vigour' or 'health'. Elsewhere, while discussing the vital part played by the release process in rehabilitation, they reinforce this confusion: '[t]he fitness of the animal and the timing, location and other circumstances of the release must be carefully considered' (p. 236). This is clearly a reference to fitness as a quality that can be assessed, and so something akin to vigour. Since conservation-based criticisms of wildlife rehabilitation frequently rest upon a misuse of the term 'fitness', they are undeniably weakened.

Kirkwood and Sainsbury's claim is further inaccurate in suggesting that the less fit are those individuals that become *sick or injured*. This is quite wrong. Fitness ought to be judged, not merely on the basis of *succumbing* to one or another ailment, but also on the basis of the ability to recover. Evolutionary fitness may be understood in one of two senses. In one understanding it is synonymous with 'survival' – that is to say, those individuals that are fit are those that survive (Darwin's 'survival of the fittest' is seen to be tautological).[1] With this reading of 'fitness' an individual can only be deemed unfit if it does not survive, and so clearly recovery (from disease, for example) is just as relevant as is resistance to disease. Alternatively, fitness may be understood as meaning 'good design' – a quality that an animal passes on to its offspring but which cannot be predicted in advance. Animals with a thick coat may turn out to be better designed for survival than those without such coats, should the climate turn colder.[2] As with the previous reading of fitness, an animal's good design is as reliant upon 'ability to recover from' every bit as much as it is upon 'resistance to'. This is so, not only because the ability to fight against adversity is a sign of good design. Many debilitated animals have succumbed, not because of a flaw in their design, but because they have simply met with misfortune; they have been unlucky. Very often, debilitated animals have been in the wrong place at the wrong time; the victims of an oil spill, for example. Darwin himself said that such individuals may be every bit as fit as other members of the species. Misfortune is unrelated to evolutionary fitness. So however 'fitness' is understood, the idea that individuals earmarked for rehabilitation can be identified as 'less fit' is ill conceived.

---

1 Arguments in support of this claim can be found in Shimony (1989) and Ollason (1991).

2 This example has been taken from Gould (1977, p.42) who is a proponent of the 'good-design' theory.

However, wildlife rehabilitation is still open to the challenge that if an animal requires *help* in order to recover, this must indicate that it was, after all, 'less fit'. The treatment of debilitated individual animals may mask the fact that they are 'less fit'. Treatment may mask one of two things. It may mask the animal's poor design (which was responsible for it succumbing in the first place) or, assuming it to be unlucky to have succumbed, it may mask its lack of fitness in terms of ability to recover on its own. Either way, natural selection would appear to be compromised.

Whilst it is important to recognise the help that rehabilitation offers debilitated wildlife, such help does not, in fact, promote the survival of less fit individuals. Rehabilitation in itself requires a certain capability for coping. There is no doubt that it can be a stressful process as Kirkwood and Sainsbury (p. 238) note:

> ... the stresses of capture from the wild, hospitalisation, treatment and subsequent release are hard to assess but may be substantial. For example, Rebar *et al* ... considered that confinement and handling may have contributed to deaths due to shock in oiled sea otters at rehabilitation centres.[3]

If an animal is to survive not only its debility but rehabilitation too, it must be an animal that is well designed for survival. A likely objection will be that fitness, as a product of *natural* selection, demands a selection procedure devoid of human influence. Alongside this may be found the related objection that an animal's capability to survive rehabilitation procedures is not a relevant capability; it is of little value in nature. The way Kirkwood and Sainsbury state their case gives weight to these objections. Natural selection is usually contrasted with artificial selection involving human determined reproductive outcomes, such as in domestication. But the contrast Kirkwood and Sainsbury are making is of a more rigorous kind, in which natural selection is taken to mean selection devoid of any human influence. Quite apart from the question of whether, given the all-pervading nature of human influence upon the natural world, it makes sense to make such a rigorous distinction (see Chapter One), natural selection is rarely understood in this way. Kirkwood and Sainsbury themselves make the implicit assumption that individual animals that do not succumb to debility (or indeed, recover on their own) are 'fit', thereby legitimising the human-influenced environment as one in which it makes sense to talk of fitness. Indeed, to do otherwise, is to invoke some kind of mythical selection procedure in which human influence does not feature. An animal's good design must relate to a real context not an illusory one, and it is a fact that wild animals' environments include humans and human hazards.

It is a separate question as to whether wild animals can become well designed in a human-influenced environment since we are uniquely random in our actions. We oscillate between destruction and protection in a most haphazard way, and, as I discussed in Chapter One, it may be that species require consistent pressures in order to become adapted. Irrespective of this, given our ubiquitous presence and widespread impact on natural selection, it would be quite wrong to single out rehabilitation as guilty of compromising natural selection and undermining

---

3 See also Schmidt (1997).

Darwinian fitness. If rehabilitation is guilty, then so too, are all our activities, including conservation – be it the management of a wood or the reintroduction of a species. It is not clear that there is a difference in kind between the conservation practice of re-establishing a threatened species (that is losing the battle in the face of human hazards) and the rehabilitation practice of aiding the recovery of a compromised individual animal. Nor is it clear that there is any important difference (in terms of 'fitness') between, say, providing red squirrels with food-hoppers (or managing a woodland for them) and rehabilitation. Indeed, since rehabilitated animals have survived debility and the process of rehabilitation, they have survived not (as may be said of red squirrels whose competitors, the greys, are controlled) by the removal of challenge, but rather, by the replacement of challenge. And as discussed already, the challenge presented by rehabilitation is no more different, or less relevant to wild survival than a plethora of other challenges faced by wild animals. Getting by in the face of direct and indirect challenges from humans is the lot of today's wild animals.

Though wildlife rehabilitation is merely one of a whole range of human influences upon natural selection, the different influences require differing evaluation. Rehabilitation is, in common with aspects of conservation practice, of a particular kind of human influence. It is reparational in nature; it is in the business of making amends. This is in contrast with many other human activities that are indifferent to, or even antagonistic towards all concerns other than human ones. Kirkwood and Sainsbury (p. 239) are quite wrong, for instance, in claiming that '[t]he treatment of compromised individuals ... is no less an interference than shooting the fit' if their intention is – as it would seem to be – to suggest that they ought to be evaluated in the same way. Their claim has to be rejected immediately on pragmatic grounds, since evolutionary fitness is not a quality that we can predict or even recognise. Shooting the fit would be an impossible task. However, putting aside this practical difficulty, there are other factors that illustrate the difference in nature between helping the less fit and shooting the fit.

An important point of distinction between the two approaches is the degree to which they are amenable to human control. Help can never dictate a desired outcome; it can merely facilitate one. Shooting, on the other hand, has a far more certain outcome. Furthermore, unlike shooting, rehabilitation requires an input from the animal itself – it is reliant upon the animal's coping mechanisms. Although an animal will sometimes manage to avoid being shot, more often, the animal's wiles and the gun are grossly mismatched.[4] Shooting generally works against the animal (in that it leaves little or no room for drawing upon survival strategies) whereas help in the form of rehabilitation works *with* the animal. In this respect, rehabilitation is shown to be much closer to Kirkwood and Sainsbury's ideal of wild animals as having 'always fought their own battles' than is shooting. Shooting leaves little room for a battle at all. So although the criticism that rehabilitation compromises natural selection has immediate – and popular – appeal, it does not stand up to scrutiny. There is no evidence to suggest that wildlife rehabilitation works against

---

4 Even where animals escape with injury, the types of injuries inflicted by the gun often preclude the possibility of recovery.

the conservation values of 'natural selection' and 'fitness' that might lead to an undermining of conservation. At least this one example of prioritising the individual is compatible – in the terms discussed – with the practice of conservation.

Nevertheless, there is no room for complacency. Concern for the individual does not compromise the particular conservation values of natural selection and fitness. But there would appear to be little room for this sentiment within the holistic philosophy adopted and practiced by conservationists. This is particularly noticeable where that holistic philosophy sanctions the practice of culling.

Conservation, culling and prioritising individuals

The culling of individual animals is a holistically driven conservation strategy. For example, in order to promote the revival, in Scotland, of the capercaillie species, the RSPB has undertaken extensive culling of foxes and crows (Linklater, 1997, p. 7). The culling of red deer in parts of Scotland is believed to be an ecological necessity, the deer population having apparently increased to the point at which they are 'out of control' and an ecological liability (Scott, 1991; p. 848).

The conservationist's pro-culling holistic argument contains three premises. The first is that conservation value is placed on the 'whole'; individuals are mere members or parts of the 'whole'. The second premise is that conservationists ought to promote the good of the 'whole', be this species, habitat or ecosystem. Given the third premise – that selective culling is good for the 'whole' – conservationists conclude that selective culling is desirable for the sake of conservation.

Conservation value is placed upon the 'whole'

If the protection of 'wholes' is of paramount importance to conservation, it is curious that culling, by its very nature, is a form of *compartmentalising* nature. To advocate culling is to differentiate, to segregate, and to divide nature into parts according to perceived worth. In the case of red deer culling, the red deer species is isolated from the wider environment that it inhabits, in its identification as an unwanted 'part'. And further, the process of culling involves differentiating between those groups of deer to be targeted and those to be left alone. The culling of red deer, for instance, is achieved by targeting hinds of particular age and condition (Wigan, 1993; pp. 27; 137). If conservationists' 'wholes' (themselves, of course, only parts) are to be holistic concepts at all, then they ought to be seen as being whole 'wholes'; that is, complete in themselves and part of the wider picture. The notion of culling seems to be anathema to a coherent approach to the protection of 'wholes'. This is not to argue that culling is a threat to the *viability* of 'wholes' (such as species) but only to claim that it is against the *spirit* of holism.

The conservation supremacy of the 'whole' must be questioned further. In advocating the sacrificing of individuals for the sake of the 'whole', conservationists usually (thankfully) exclude human individuals. They cannot, therefore, be taken *literally* when they argue that the 'whole' is always of greatest value. This must be seen for what it is: mere 'conservation-speak', rough shorthand that suggests where their priorities tend to lie. Humans, after all, are the worst offenders of environmental damage. It behoves the conservationist to explain why the taking of non-human life is to be so very differently regarded to the taking of human life. Yet even assuming a convincing argument could be made, the conservationist will still surely find it difficult to justify the taking of non-human lives for the sake of 'wholes'. The fact is that the value of these 'wholes' is not only usurped by the value of human lives, but also by very much more trivial things to do with human comfort, amusement and convenience. Though individual conservationists may spurn certain conveniences, conservationists do not, as a body, boycott cars, meat eating, or houses without cavity-wall insulation or solar heating, all of which contribute to the degradation of the environment. Nature's 'wholes' may be worthy of protection but there are clear limits to the costs involved. These limits cannot merely be *assumed* by conservationists to include the culling of non-human individuals. Indeed, the difficulty is in understanding how conservationists can justify culling. That is to say, given that protection of the environment is so readily usurped by trivialities, it is difficult to accept that animals' lives may be legitimately sacrificed for the sake of environmental protection. Furthermore, the argument that humans ought to be afforded special treatment because they are not part of nature is not in tune with a holistic philosophy. Nor does it sit entirely comfortably with the conservationist's championing of nature.

In the conservation clamour to protect 'wholes', not only is the necessity for particular measures often taken for granted, but so too may the true nature of such measures be masked. Such is the case with culling. We should not forget that the term 'culling' is a euphemism for 'killing'. Too easily a subconscious connection is forged between conservation and culling, as if they somehow belong together. This is facilitated, no doubt, by the apparent innocuousness of the term 'culling'. It has become sanitised killing – part of the great conservation quest. But the act of killing de-humanises us and takes its toll, even on the most hardened of killers. Wigan (1993, p. 96) in discussing a particular red deer cull in Scotland in which the population is regularly reduced by 35 per cent writes:

> Something that is seldom taken into account is the emotional stress to stalkers having to shoot animals in these numbers, piled upon the physical stress. Only those who have done it, who have followed up an unmothered calf, hanging round its suddenly-deceased parent, yet somehow awkward to get a sure shot at, will understand the stress I am talking about ... No one normal enjoys shooting dependent calves, or a slaughter of hinds. A large estate cull in east Sutherland had to be interrupted for a week while stalkers recharged their psychological batteries before continuing.

Lacy (1995) makes culling sound more acceptable by characterising it thus: 'Culling is the termination of the life of an animal before it would have died from unavoidable disease or failures of organ systems (old age or natural causes)' (p. 118). It is as if culling were merely the end-result that comes about anyway. But such consequentialist thinking misses a vitally important point. To view killing in this way is to ignore the importance of process: the process of life and the process of death. Neither the wonder of life nor the way in which death comes about is seen worthy of mention. Yet for those conservationists (such as Lacy) who adhere to an evolutionary biological approach to conservation, process clearly does matter. And if process is of relevance to conservation, then it is not self-evident that the processes surrounding individual lives are of less importance than the processes connected with species. Indeed, to dismiss the former is to impoverish conservation. The wonder of lived lives provides most of the motivation and energy that is required for the uphill task of conservation. Deer watching, for instance, inspires many, and in Scotland, the red deer has become a national symbol (Scott, 1991; p. 848). Recognition of the value – to conservation – of the processes connected with individual lives prompts a quite different perception of culling.

Conservationists ought to promote the good of the 'whole'

In order for conservationists to promote the good of the 'whole' they must first recognise which 'wholes' are worthy of this help and how to assess their relative value. In any one area there will be a number of 'wholes' that might be considered worth protecting (a variety of habitats, species, subspecies or populations). With no objective decision-making process, judgments are made according to the particular perspective of the conservationist in charge. But not only are judgments very much dictated by individual perspectives, environmental policy decisions may themselves add to the often-arbitrary nature of environmental protection measures. For example, Deakin (1997, p. 70) points out that '[t]he designation of an SSSI often dictates a tunnel vision of the habitat, skewed towards the debatable selection of what is deemed "scientifically interesting"'. He discusses the Countryside Council for Wales' decision to remove more than 900m of hedgerow in North Wales in order to encourage waders to feed there. Talking of the terrible dilemma between the uprooting of fast-disappearing hedgerow and the creation of open space in which declining lapwing thrive, he writes that:

> Uprooting hedgerows is like taking an India rubber to the history that is written into the landscape. It also silences what Richard Mabey memorably calls "the rich harvest of bird song". Of course, saltmarshes and wetlands are valuable habitats – and beautiful places – but who is to say that a dunlin is more "valuable" than a dunnock? (p. 70).

Even within a single organisation, the task of prioritising is problematic. Wigan tells of the confusion over the Nature Conservancy Council for Scotland's management of SSSIs:

> It often seemed to the ordinary farmer, crofter or landowner that within the conservancy one hand did not know what the other was doing. The peat-bog specialist would arrive in nesting-time to trample over a site just described as vital for rare nesting birds (p. 77).

In parts of Scotland, red deer populations are considered to be too large and culling seen to be a necessity for two distinct but interconnected reasons. The first is the requirement for the regeneration of past habitats (usually woodland) and the second, the desire for protection from degradation of existing valued habitats (again, usually woodland). Additionally, it is sometimes argued that, in the absence of predators, culling is necessary in order to prevent deer starving to death. This latter point I will deal with later.

It is supposed, usually without critical discussion, that tree regeneration is a laudable conservation objective. However, it is not obvious that the forested hillside has greater conservation value than the treeless hillside. Each habitat has its own combination of plant and animal species, its own ecological interest and attractions. As discussed in Chapter Four, the mere existence of a past habitat (the fact that Scottish hills were once bedecked with trees) cannot provide legitimate reason for its attempted recreation. The 'thinking interventionist' would be required to take account, not only of raw nature, but also of the effect of the more recent past in which Scottish hills have become known and cherished as treeless habitats. Indeed, '[i]t has been proved conclusively that, from a scenic viewpoint, the Highlands are loved for their openness and space' (Wigan, p. 83). As an activity rooted in the concept of care, and as a human enterprise, conservation ought to embrace such deep feelings about landscape. Other factors, too, need to be taken into consideration. Recalling the arguments of Chapter Four, the 'thinking interventionist' may not, for instance, be able to sanction change that necessitates the intensive intervention that culling requires.

To justify woodland regeneration conservationists often adopt the argument of historical precedence. Yet this argument is every bit as susceptible to bias as is the evaluation of different 'wholes'. There is no objective way of deciding which period from the past should be emulated. Although evidence from the pollen record informs us that most of Britain was covered with forest around 7,500 years ago (Yalden, 1982, p. 10) the fortunes of forests have varied considerably over time.[5] Earlier, from at least 70,000 to 9,000 years ago tundra conditions prevailed throughout Britain (Dyer, 1990, p. 23), while later, forest cover waxed and waned according to climate, human activity and disease (pp. 24–63; 92–118). From about 5,000 years ago 'the representation of tree pollens becomes, erratically but inexorably, less and less [with the] minimal ... reached around 300 bp' followed by evidence of replanting (Yalden, pp. 9–10).

Conservationists may argue that there is no bias involved, that by removing the deer they are simply allowing the land to do what it does best in the present climate – which is growing trees. The problem with this argument is that it relies upon an assumption that the appropriate circumstances for allowing the land to do what it is good at call for an absence of deer. Yet it is not clear why this should be so. There seems no reason – other than bias – why the presence of deer ought to be any less acceptable a circumstance than the absence of deer. It could be argued that, if left to its own devices, the land is good at supporting deer. Perhaps, though, the conservationist equates large deer populations with human interference and so with circumstances that run contrary to a conservation rationale. But such reasoning would be difficult to sustain. Not only is human influence all pervading – such that it is unreasonable to characterise conservation as separate from it – but conservation makes use of human intervention for its own ends. Consider, for instance, the re-introduction of species, the re-flooding of wetlands, and the use of grazing animals to maintain certain habitats. Conservationists may view human impact positively or negatively. A clear example of negative impact is the destruction of a habitat (such

---

5 See Yalden's (1982) pollen diagram, p.6.

as has recently occurred with the building of another runway at Manchester Airport). However, when a habitat is changed from one type to another, there is no obvious sense of conservation damage and it may simply be bias that dictates the acceptability of one but not the other.

If the arguments of historical precedence or of allowing the land to do what it does best cannot convince us that tree-covered hillsides are of greater conservation value than deer-covered hillsides, the argument for biodiversity does little better. The conservation call to maximise biodiversity is a call for maximising representatives of as many different 'wholes' as possible. But as we have seen, there are no satisfactory answers to the question of how to compare or evaluate different 'wholes'. While wooded hillsides carry representatives of different species of tree and support, for example, red squirrel, goshawk, redpoll, capercaillie and Scottish crossbill, open hillside will be covered in plants such as heather and grasses, and support species such as golden eagle, merlin, short-eared owl, ptarmigan, meadow pipit and mountain hare. A number of moorland and grassland species are declining, in part because these habitats are themselves declining. For example, the heather-covered moors, upon which the merlin depends, are 'under threat from forestry planting or conversion to grass through sheep grazing or re-seeding' (Cady & Hume, 1992, p. 136). Such habitat destruction 'is almost certainly the major cause of [the merlin's] decline'. Wigan (1993), too, observes that under-grazing can disadvantage moorland wildlife (p. 95). Similarly, the golden eagle, arguably Scotland's most celebrated bird, relies upon moorland and bare hillside for its hunting. Therein, it seems, 'lies one of the main threats to its future, for such areas are becoming the target for forestry plantations' (Cady & Hume, p. 124). Two other moorland birds that have declined as a result of the spread of large conifer plantations are the raven and the skylark. Skylarks 'have probably benefited more than any other birds from the advent of arable farming and the deforestation of the countryside' (p. 208). To argue, then, on the grounds of biodiversity, that forestation is always a conservation good is clearly mistaken.

In assessing biodiversity, the boundaries around 'wholes' are not clear-cut. Considering how to evaluate the biodiversity value of Highland forests, it is not clear whether these areas ought to be evaluated with respect to the rest of the Scottish Highlands, Britain, Europe, in a global sense, or any category in between. Assuming a very parochial evaluation (as biodiversity is often considered important at the local level) it is not obvious that the biodiversity of the Scottish Highlands would be improved with more wooded areas. There are already several examples of extensive forest such as Glen Affric. And assuming a wider evaluation, it may be that the Scottish treeless hillside – particularly in combination with the red deer species – offers the most in terms of biodiversity. Wigan notes that the bare-hill-with-deer combination offers a unique kind of stalking habitat, sought after by those from abroad (p. 134). It is feasible that this habitat might be valued, instead, for its contribution to biodiversity, and for deer watching rather than deer shooting. As a conservation objective, tree regeneration requires greater scrutiny. Woodland cannot be assumed, uncritically, to be a more desirable conservation objective than open-hill habitat.

While one reason that conservationists advocate red deer culling is to allow for woodland regeneration, the other is to protect existing (usually woodland) habitats.

My arguments relating to woodland regeneration apply just as readily here. Arguments for historical precedence, the land's 'natural ability' and biodiversity fail to lead to any conclusions about what kind of habitat is – from a conservation perspective – most valuable. Historical precedence and the attendant human attachment to habitat may sometimes indicate preservation (in the sense of safeguarding) as the best conservation strategy. However, it may not do if, for example, strength of feeling raised by plans to cull deer is weighed alongside. Furthermore, where attempts to honour concerns such as historical precedence and human attachment to habitat necessitate intervention of an extreme kind – as would be rejected by the 'thinking interventionist' – then conservation is better served by non-preservation.

Surely, though, preservation is what conservation is *about*. Is it not the case that the removal of preservation from the conservationist's remit removes the very backbone of conservation? In Chapter One, I argued that while the protection of threatened species is often the conservationist's greatest concern, this can mask the greater importance of protecting nature's ability to continue in the time-honoured way: change through the process of speciation. This is the true purpose of conservation, and preservation must ultimately represent this goal. Preserving diminished habitats is a laudable conservation strategy. But, contrary to the conservation rationale I have proposed, it is to be understood as playing a bit part rather than the leading role. As such, it must take its place alongside other conservation goals. All things being equal, the protection of threatened habitats ought to be implemented, but habitat preservation should be seen, now, as one conservation strategy among many. It is only one of a range of concerns, rather than conservation's ultimate goal. When preservation threatens other conservation strategies it needs to be reappraised. Conservation may sometimes be better served by allowing habitat to change. This may be so where, for example, the preservation of a habitat or species necessitates activities – such as culling – that are anathema to a caring attitude. Or it may arise where intervention leaves no room for future nature, such that nature is preserved as if it were an ark. Provided change allows for the possibility of evolution, it will satisfy conservation's ultimate goal.

Clearly, then, conservation can allow only change of a particular kind. It is restricted not only by the need to honour both a caring attitude and the unfolding of future nature, but above all by the need to allow for the possibility of speciation. Conservation can never, then, sanction the replacement of habitat with something else, with something *different in kind*, such as a road or a car park. However, change in the form of one habitat evolving into another habitat may be quite acceptable. The human activity of conservation is both guided and buffeted by the need for an expression of care, a demand for the unfolding of future nature and a requirement for the possibility of speciation. A demand for conservation will always be a demand for the continued opportunity for speciation. It will, however, only sometimes be a demand to preserve a habitat *in situ*. Conservation should not, of necessity, require that existing habitats – even those genuinely representing threatened habitat types – be protected in a way that preserves their current character. With this in mind, the conservationists' argument for deer culling is still further weakened.

It is still very much the case that conservation is needed in order to protect habitat from degradation, but I have redefined the notion of habitat degradation. A degraded habitat is one that can no longer offer the possibility of a range of speciation processes.[6] Conservationists cannot reasonably argue that the change from woodland habitat to moorland or grassland habitat is an example of habitat *degradation*. Even if it were agreed that *permanent* disappearance represented either loss or degradation, the impact that deer have upon woodland could not be so described. For instance, in a discussion of one man's rigorous deer culling policy, Wigan notes that as soon as the grazing deer had gone, saplings sprouted up vigorously, '[p]roof, perhaps, that the notorious degradation of habitat is an over-simplification, or a misrepresentation' (pp. 80–81). That is, although the presence of deer had prevented tree regeneration, the possibility of such regeneration was, so to speak, waiting in the wings. Even where so-called erosion occurs, 'normally the herbage pattern changes and coarser grasses form a healing carapace' (Wigan, p. 95). Yet to emphasise the likelihood of erosion *healing*, is arguably to regard it in the wrong way. Erosion is part of the natural process, and just as it closes off certain possibilities, so it creates others. Destruction in nature, by nature, ought to be viewed positively. Tickell (1995, p. 27), for instance, notes the importance to conservation of large, destructive wild animals. Indeed, the traditional line that large populations of deer cause habitat damage through overgrazing is oversimplified. There is evidence that a *dearth* of deer can have a similar effect. Wigan acknowledges that an 'over-pressurised' hill is 'disadvantaging other moorland wildlife being unsightly, and resulting in the ground taking a while to recover' (p. 95). But what is interesting is his comment that:

> [a]ll of these things are true of a less-often mentioned phenomenon – under-grazing, now occurring on deer range that for a variety of reasons has been vacated by deer.

With the acknowledgment that grazing in itself is not necessarily a conservation harm[7] comes the realisation that popular notions of environmental destruction are misguided. It simply cannot be the case that grazed land is of necessity degraded habitat, or even (conservation-wise) poorer than non-grazed habitat. It can be said, only, that grazing sometimes thwarts the individual aims of particular conservationists. Contrary to the tone taken by some conservationists, the red deer cannot be branded, *per se*, as a villain.

It is a fact that nature is as much concerned with destruction as construction; they are both essential processes in every ecosystem. However, to acknowledge the importance of death and destruction is not to compromise the argument against culling. To imagine that culling is legitimised as conservation practice because death and destruction are essential natural processes is to fall into the trap in which

---

6 As was argued in Chapters One and Six, it is not merely speciation per se that is required by conservation, it is a range of types of speciation, including both the rapid speciation of, say, insects, and the slower, more deliberative speciation typical of large mammals.

7 Wigan notes that grazing is used in some areas to improve habitat and aid in, for example, heather regeneration (pp.105; 107).

Budiansky (1995) has found himself. As I mentioned in Chapter Five, Budiansky argues that since natural forces sometimes destroy forests, destruction by humans is, if not obligatory, then at least desirable. But we are not, in our conservation efforts, obliged to imitate nature at all. Conservation is not a copying exercise; it is a human enterprise. Certainly, we must look to nature for guidance, as a template for our action, in order to inspire us as well as curb our excesses. But it has never been – nor should it ever be – the remit of conservation to uncritically follow nature. Nature, after all, brings us droughts, plagues, diseases, tornadoes and the like, and it is clearly not part of conservation's remit to attempt to mimic these things.

Budiansky's argument may be that natural processes ought to be mimicked if the end product is desired. For instance, since forests benefit from regeneration after storm damage, we should mimic this by clear-cutting. However, Budiansky fails to address the fact that the two processes are entirely different. After storm damage, fallen trees are an important element of the regeneration process, serving, amongst other things, as shelter, frost-protection and nutrient for the emerging seedlings (Russell, 1997). Clear-cutting – where trees are felled and removed – does not mimic this set of circumstances. There is the mistaken assumption that destruction is destruction however it comes about. But each form of destruction – no less than each form of construction – involves its own idiosyncratic processes. It is mistaken, then, to base an argument for a particular conservation action (e.g. clear-felling) on the value of a natural end-product (e.g. post-storm forest regeneration) when quite apart from the differences between human agency and the natural occurrence, there is so much discrepancy between the two processes.

Clear-cutting does not adequately mimic natural forest damage. Nor does culling adequately mimic natural die-off. Considering the example of red deer, natural death is never – as it is with culling – restricted to a particular sex or age ratio within the herd. And natural death will bear a much greater relation to illness and old age than does death through culling. Furthermore, there is a real relationship between natural die-off and carrying capacity. In contrast, the relationship between deer death through culling and carrying capacity is artificial. This is so in that either carrying capacity can only be guessed at, or it is determined with particular parameters in mind (such as the carrying capacity for habitat x that will allow for woodland regeneration).

Justification, then, for the value of culling as a conservation exercise, is singularly lacking. Since conservation does not, after all, demand forest (or any other kind of habitat) regeneration, deer culling for this reason cannot be sanctioned. The importance of preservation lies not in the maintenance of unchanging habitats but in protection of the opportunity for speciation. In this light, culling serves no conservation purpose. Nor can culling gain credibility as being equivalent to the death and destruction to be found in nature. Conservation is a thinking, caring, human enterprise and not merely an attempt to mimic nature. If the culling of red deer has lost credibility as a conservation measure, then it follows that there should be less reason for antipathy towards the rehabilitation of members of this species on conservation grounds.

Selective culling is good for the 'whole'

It may be objected that the claim that any habitat is as good as any other is too extreme. There may, sometimes, be good reasons for wishing to conserve particular habitats, and in order to achieve this, culling particular species may seem a reasonable strategy. Yet even when a specific end-goal is desired – let us say, the protection of a forested hillside in the Scottish Highlands – the value of culling is not as clear-cut as might be supposed.

Let us accept – for it seems to be true – that in deer populated areas, without intervention of some kind woodland regeneration will not occur. Accepting, for the sake of argument, that woodland has greater conservation value than moor or grassland, it would appear sensible to reduce deer numbers through culling. But there are two problems with this apparently simple solution.

The first is that it fails to take into account the fact that deer population is just one factor out of many[8] that dictates the fate of a habitat. Of particular note is the effect of sheep grazing on habitats. Though aerial photographs have provided evidence 'that Scotland's vegetation profile is changing' the causes have never been established. The stark fact is that 'sheep effects were never separated out from those of deer or climate', and the:

> [d]eer population reduction has seldom been talked of in the only sensible way – in conjunction with reduction in sheep numbers ... [r]educing one and not the other will merely lead to increased fertility, bodily condition and productivity in the other (Wigan, p. 142).

Indeed, the big increase in red deer, which took place between 1965 and 1975 (after which numbers increased more slowly) appears to have occurred as a result of sheep stocks falling by 20 per cent between 1966 and 1978.[9] Where habitat protection is sought, deer culling is not the panacea it is often supposed to be.

The second problem with taking the simplistic deer-culling line is that, according to Wigan (p. 143), in order to achieve woodland regeneration deer need to be virtually (or actually) eliminated. But when a deer populated area has all of its deer removed, there occurs the 'sump effect' whereby deer are sucked in to ungrazed ground by the more succulent feeding it offers (p. 82). Wigan notes that the East Grampians Deer Management Group 'submitted that to create a deer-free zone on an area of 50,000 hectares of open hill would require a reduction in numbers over an area ten times that size'. It seems, in fact, that '[i]t will never be possible to kill out deer in a localised way and expect the gap to remain empty'.

The upshot of the problem with creating grazed-free zones is that if woodland regeneration is a serious aim, there is no alternative but to employ fencing. Trying to achieve natural regeneration without fencing is, in deer expert Ronnie Rose' opinion, 'like making a circle in the sea and trying to stop the water coming in'

---

[8] Other factors include climate, human impact, and the effects of other grazers such as rabbits.

[9] Wigan explains that the biomass 'shifted from sheep to deer' (p.141).

(Wigan, p. 76). Though fencing can be effective in allowing regeneration to occur (Tickell, 1995, p. 27), some conservationists object on the grounds of artificiality. The absurdity of this complaint, coming as it does from the same conservationists who advocate culling, is quite profound. And further, if fencing is rejected in favour of culling alone, in order to control the deer it will be necessary to create artificial clearings. So to insist on culling but not fencing on the grounds of artificiality is perverse. At the moment, culling is employed alongside fencing and it may be that fencing on its own would not achieve the desired results. However, given that fencing is, as it were, the active ingredient in the recipe for regeneration it would not be surprising if, with a little thought and ingenuity, it could be effective without the need to cull. It might even be that fencing could be dispensed with. In imaginative and forward-thinking experiments, lion-dung has been used successfully as a deer-deterrent.[10] But whether fencing is used or not, culling is not the answer to the so-called conservation problem of over-population.

There are problems, then, with all three of the conservationist's premises that lead to the conclusion that selective culling is desirable for the sake of conservation. We can therefore legitimately reject this conclusion. Nevertheless, a difficulty yet remains: what of the apparent clash of interests between the conservation practice of culling and wildlife rehabilitation? Before discussing the implications of my arguments concerning culling, I will make two general points about the current relationship between culling and rehabilitation.

Firstly, where there is a perceived need for deer culling, it is considered necessary on an area-by-area basis rather than as a blanket prescription. As Wigan observes:

> The desirable density of grazing animals will differ from area to area ... recognition is now universal, amongst the bodies and individuals that count, that deer management should be addressed on an area by area basis. An obvious illustration: at present over half the wild cull is from the Grampians, a small part of deer range geographically. In other parts of the Highlands numbers probably could never have built up to support such a cull ... The carrying capacity of hill ground varies widely (p. 142).

Since deer culling as a conservation requirement is situation-dependent, there is no logical clash between the practices of culling and wildlife rehabilitation. There is no contradiction in conservationists adopting culling programmes with some deer populations and rehabilitation programmes with others.

Secondly, and more specifically, the majority of culling is ostensibly employed to reduce numbers rather than to achieve extermination, and in this conservationists enjoy a degree of public support. Whether conservationists would continue to enjoy the current level of public support for their killing of animals were they to admit more freely to the actual extent of some of the killing (sometimes near extermination) is very questionable. Wigan discusses the radical reduction of deer employed by Dick Balharry, manager of Creag Meagaidh, 'a 4000–hectare former deer forest [where d]eer numbers between 1977 and 1990 were reduced by 91 per

---

[10] Northwest Tonight, 1992.

cent' (p.80). He observes that Balharry 'is reluctant to focus too hard on how many deer have had to be sacrificed for his project. The counting he prefers is of trees' (p. 82). He infers that public approval would decline 'if the rest of the Highlands was a thicket of young trees' (p. 83).

By strict definition, to cull (to select) is to control rather than to exterminate a population. Culling, as a conservation exercise, arguably depends for its level of public approval on this definition being put into practice. After all, the red deer 'has become synonymous with the romantic grandeur of Nature' (p. 126) and in Scotland is something of a national symbol. As an exercise in maintaining population 'balance' culling does not clash with the practice of wildlife rehabilitation. Since the ethos behind culling is not 'the less the better', there is no logical difficulty with having both practices operating side by side. Though rehabilitation would not sit comfortably with an extermination programme, conservationists may well be on shaky ground in employing such methods anyway.

Even though there is no logical incompatibility between deer culling and deer rehabilitation there is clearly a theoretical tension in the sense that each practice rests upon a very different set of assumptions. While wildlife rehabilitation takes the plight of the individual seriously, the practice of culling sweeps aside any value the individual may have in pursuit of a holistic goal of some kind. In my earlier discussion about the difficulties with the holistic position, I concluded that conservation ought not to be driven exclusively, or with too much passion, by holistic goals. In particular, the arguments in support of culling were entirely unsatisfactory. Through scrutiny of the red deer situation in Scotland, I can now say that the arguments that favour deer culling do not stand up to scrutiny. This supports my general claim that culling for the sake of conservation is mistaken. Specifically, the conservation goal of woodland regeneration is wanting; there is insufficient evidence that it is a valid conservation strategy. In addition, I found deer culling to be ineffective as a means of achieving woodland regeneration, even if this were a valid conservation aim. Deer culling, for the sake of conservation, is therefore unjustified. The main conservation objection to the rehabilitation of red deer was that it is at best pointless, or at worst harmful, in working against conservation's culling policy. Since this objection is no longer meaningful (and since the objection relating to interference with 'fitness' has been quashed) there seems no reasonable barrier preventing the conservationist from embracing the concept of red deer rehabilitation.

However, one final worry remains. Culling will sometimes be advocated for the sake of individual animals. Wildlife rehabilitators – who have the welfare of the individuals at heart – should, it may be argued, view culling favourably, where it prevents slow death through starvation in those cases where numbers have escalated. That is, culling may be seen to have a humane function. Furthermore, in such circumstances, wildlife rehabilitation may seem to be contra-indicated. That is, it may be thought illogical to return a rehabilitated deer to the wild if food is scarce and the deer population is thought to exceed carrying capacity. The rehabilitator's own moral code – care for the individual – may, it seems, sometimes prohibit rehabilitation practice being carried out.

Culling and individual welfare

According to Wigan, the issue of escalating deer numbers has been exaggerated and misrepresented (p. 141). He writes that 'the high-profile drive against deer numbers ... fell victim to its own propaganda, and eventually deployed blatant misinformation to advance its case'. Part of the misinformation relates to the negative impact of the increasing populations upon the individual deer. It seems that with the increase of deer in the 1970s 'declining bodily condition and increased mortality due to population density never occurred'. However, if Wigan is right, then what is to be made of comments such as those of Watson (1989, p. 24):

> The effect of over population on the animals themselves is, by any standards, unacceptable. Red Deer on the open Scottish hills are, even under good management, stunted specimens of their race. The lack of winter fodder that results from year-round heavy grazing can reduce them to starvation in the late winter and early spring, or at least bring them to a state of malnutrition that inflicts lengthy suffering and leaves the animals highly vulnerable to the stress of hard weather.

It cannot be denied that some years, large numbers of deer die from starvation. Wigan himself admits this (p. 100). Talking of the winter of 1992–3 – at the time of his writing – he says that when the records are collated, deaths are certain to be high. However, the reason for the high mortality, Wigan claims, is climatic. A lack of rain in May 1992 produced little summer grass for fattening. Summer was followed by prolonged, inhospitable rain through to January 1993. He goes on to say:

> Those who hasten to attribute deer deaths in 1992–3 to overpopulation should think again: deer deaths were as numerous in lowly-populated areas as in densely-populated ones. The explanation lies in the weather cycle. Nor is it simply to do with the fact of rain. It is when the rain falls, or fails to fall, that matters. The December/January Highland rains, which can be heavy, do not affect deer unduly. But if they enter the winter in poor condition it is a long time to wait until first grass in April or May from which to build up recovery (p. 101).

Talking of the many deer deaths in the winter of 1989–90, the pertinent point revealed is this: 'deaths had occurred in places where there was no perceptible population problem [such that i]t is now accepted that the relationship between population and winter death is not simple' (p. 93). Indeed, work by scientists on Rum's deer populations has shown that 'even heavily culled populations will not always affect the proportional number of winter deaths' (p. 17). The grounds, therefore, 'for reducing deer numbers for humane reasons, to prevent them dying in bad winters, have weakened'.

It is not, presumably, that there is *no* connection between food availability and mortality rates; rather, there are other factors that more directly influence death rates. Since calving rates change with population pressure (presumably, at least in part as a result of food availability), then it must be assumed that deer numbers will reflect the grazing quality of the previous year(s) and its potential for the coming year. However, if other factors (notably climate) conspire against the deer in a particular year, they will, as it were, be caught short, and many deaths will occur.

The reason why culling will not alter the course of events significantly, is, presumably, because of timing. Culling (which targets hinds) takes place between late October and February. By this time, though, summer fattening has either been successful or not (depending upon weather patterns) and so the fate of the deer is partly sealed. (Perhaps if culling took place in the summer it would better influence winter deaths – but of course this would be considered inhumane as calving is taking place at this time). The effects of weather, the other factor responsible for winter deaths is not, of course, influenced by culling. Weather acts upon individuals equally and not populations proportionally. It may be that there would be some benefit (after a cull) to the remaining deer the following summer if grazing were poor then. But this is impossible to predict – deer may have been culled needlessly.

The argument for culling as a humane measure is flawed because winter mortality is such a complex matter. But the argument is also questionable in its assumption that it is preferable – from a deer's point of view – to die by the gun than to die of natural causes. A strong case for the greater acceptability of a natural death over death through culling has been put for the elephant (Guthrie, 1992). According to Kiley-Worthington (1997), elephant culling cannot be achieved humanely. With any social species the distress caused to other members of the group when an individual is killed, can be considerable. Like the elephant, the red deer is a highly social species. There is evidence, too, that from the individual animal's point of view culling is a less acceptable death than a naturally occurring one, and that this may apply to deer.

A scientific study – commissioned by the National Trust to determine levels of suffering in hunted stags – revealed that stags hunted with dogs suffer enormous stress (Green, 1997, p. 63). This comes as no great surprise: common sense made us privy to this long before science was called upon. However, the science has revealed that hunted animals suffer stress at least as great as the stress suffered in road accidents. Again, this may not be terribly surprising since a hunt is a prolonged affair. This might lead us to suppose that the length of time it takes a deer to die is the significant factor in determining the degree of stress it experiences. It might be assumed that it is kinder to kill a deer quickly by shooting than to let it die more slowly of cold or hunger. But this is arguably the wrong way to look at it. Two zoologists, Chris Barnard and Jane Hurst, both of Nottingham University, have recently put forward a new theory of animal suffering. They claim that suffering is triggered 'when the world frustrates an animal's adaptive drives' (Vines, 1997, p. 31). If evolution has designed an animal to deal with a certain condition, then despite the way it may look to us, the animal may not be suffering, or at least may be suffering less than we imagine. Although this cannot be supported as a way of defining suffering,[11] it seems to add important vision to the animal welfare debate. According to this theory, the stag suffers so excessively in being hunted by hounds because it has not evolved to cope with a prolonged chase. This thinking would be in accord with the work on deer stress, set out in the *Bateson Report* (Bateson, 1997), which has found that red deer are, in fact, sedentary and relatively unfit. It is not only that British red deer have no current predator (other than humans); it is, more

---

[11] See, for example, the objections put forward by Manser (Vines, 1997, p.33).

importantly, that there was no ancestral predator that subjected them to a lengthy chase (a wolf chase, for instance, was brief).

Given this way of understanding suffering, it is no longer safe to assume that it is kinder to kill a deer quickly than to allow it to die of natural causes. It is part of the deer's long history to have been faced with inclement weather and starvation. It is possible that because these factors are part of the deer's evolution, they cause individuals less suffering than is supposed. If Barnard and Hurst are right, an animal will suffer if it is deprived of making its own evolutionary-based choices – whether these are beneficial or harmful. As Vines (1997, p. 33) reports, Barnard claims that the paradoxical implication of the theory is 'that cost itself may not have welfare significance, but frustrating the animal's ability to incur a cost may have'. So it might be ventured that – as is the case with elephants – a natural death is a better option for the deer than is culling. Culling involves the instant break-up of social groups, a certain number of injuries that result in painful, protracted death (Bateson, 1997), and the depriving of any opportunity for evolutionary choice. Assessment of the suffering experienced by other creatures will no doubt always be a thorny issue.[12] It is fair to say, though, that there is some evidence for the belief that, for the deer, death by natural causes may be preferable to death by shooting. Certainly, there is no clear evidence that even if deer mortality were proportionally related to population size (which, I argue, is not the case) deer culling would be a welfare requirement. There is nothing, then, to suggest that the wildlife rehabilitator need view culling in a favourable light. The worry that it may be necessary to impose limits on rehabilitation practice in order to safeguard deer welfare can be dismissed.

In arguing that wildlife rehabilitation ought to be considered a legitimate and worthwhile conservation strategy, two worries arose. Firstly, there was the worry that rehabilitation would compromise the conservation values of natural selection and fitness. And secondly, there was the worry that concern for the individual could not be reconciled with the holistically driven conservation strategy of culling. We have found the first worry to be unfounded. Once natural selection and evolutionary fitness are properly understood, it becomes clear that helping disadvantaged wild animals does not undermine these values. It has also become apparent that there are problems with each of the conservationist's (holistic) premises that conclude selective culling is desirable for the sake of the 'whole'. The conservationists' argument is therefore unsound: it does not justify culling. Since conservationists will be required to rethink their position on culling, there is no longer any need to worry about discord between this conservation strategy and wildlife rehabilitation.

If conservation's holistic values provide insufficient justification for the practicing of culling, it might be wondered whether culling could ever be justified. That is, if 'the protection of wholes' is too problematic a notion to rest 'the taking of life' upon, might there be other, more convincing conservation reasons for sanctioning culling? The main thrust of conservation is the safeguarding of nature as the nature we value and cherish. The chief threat to this is the loss of nature's ability to perpetuate variety. If there were a circumstance in which culling was necessary to prevent this loss then culling could feasibly be justified. The difficulty, though, is in

---

12 For a thorough analysis of the problem see Mason & Mendle, 1993.

imagining a case where this condition would pertain. One possible example may be where one or other species introduced to an island is responsible for, or threatens, the loss of all existing species (and so the potential for future variety based on those forms).[13] However, if culling were restricted to only such cases, it would be a rarely performed conservation activity and not, as it currently is, standard conservation practice. Yet even in cases such as these, it could only be argued that culling were necessary if, firstly, it could be proven to be effective (which is often not the case) and secondly, there were no other means by which the animals could be controlled or removed (which is unlikely).

To take a recent case, Scottish Natural Heritage (SNH) recently decided to reduce the hedgehog population on the Western Isles. Introduced in 1974, the hedgehogs are now considered a threat to ground-nesting birds such as dunlin, redshank and lapwing, as they prey on their eggs. Initial reports claimed that SNH planned to trap and kill 5,000 hedgehogs (Kelbie, 2002). However, after a small number were killed in Spring 2003, rehabilitation experts moved in, trapping and successfully relocating a much larger number. Relocation is one possible way of dealing with clashes between wild species, but it is not the only option. Massei *et al* (2002) found that conditioned taste aversion 'is a potentially powerful, safe, and environmentally sensitive approach to reducing nest predation' (p. 1134). Their studies with laboratory rats found that 'conditioning induced a strong aversion to eggs'. With a little imagination SNH may have been able to adapt such studies to the hedgehogs in the Western Isles. And although in its infancy, there may soon be possibilities for using immunocontraception vaccines to control fertility in wild animals for between one and three years (Lowell *et al*, 1998). There are, then, very often, other ways of dealing with clashes between species. So if culling were ever justified, it would be on very rare occasions only.

Although we may not be able to talk of complete harmony between conservation practice and wildlife rehabilitation, the main areas of potential disharmony have been addressed and dismissed. Within the limits of our discussion, then, there is no obvious clash between the individualistic stance of wildlife rehabilitation and conservation practice. On the very rare occasion when culling may be a legitimate conservation strategy, the chances of it being an effective strategy as well as the only means of controlling or removing a species are slim indeed. There would seem, then, to be very little standing in the way of the conservationists' acceptance of the practice of wildlife rehabilitation. Concern for the individual, need not, it seems, compromise conservation. Indeed, these arguments, taken in conjunction with the argument in Chapter Seven that wildlife rehabilitation has positive conservation value, establish a fairly robust position. The practices of wildlife rehabilitation and conservation, far from being at loggerheads, share much common ground.

Nevertheless, convincing as this is, worries may still remain about the compatibility of wildlife rehabilitation and conservation. It would be understandable if these worries arose as a result of the press coverage that surrounds the alleged poor survival rate of rehabilitated oiled birds. Some argue that since a great deal of

---

[13] Introduced birds (over 50 species) are suggested to be one of the causes of the decline and extinction of the native birds on the Hawaiian islands (Caughley & Sinclair, 1994, p. 263).

money and expertise is required in the treatment and rehabilitation of oiled birds – most of which fail to survive in the wild – and since the number dealt with is such a tiny proportion of the wild population anyway, these resources would be better used for serious conservation projects. If a convincing case for the compatibility and complementarity of wildlife rehabilitation and conservation is to be made, this specific worry must be addressed.

Case Study: the rehabilitation of oiled birds

On the face of it, the rehabilitation of oiled birds may appear to have no conservation value at all. Indeed, it may seem to use up valuable resources that could be ploughed into protection of habitats, prevention of damage, and so on. But when looked at carefully, this first impression is found to be misleading.

The rehabilitation of oiled birds is particularly problematic for a number of reasons. Firstly, the (mostly) marine habitat in which oiling occurs is a difficult habitat from which to rescue debilitated birds. Secondly, as is the case when any wild animal is handled, the rescue operation itself causes a degree of stress. Discussing survival of sea otters after the Exxon Valdez disaster, Sharp (1996) quotes Estes (1992) as claiming that 'rescue efforts themselves caused a stress-induced mortality rate of 5–10%' (p. 226).

A third difficulty is that the harmful effects of oil often remain hidden. After the Milford Haven disaster (South Wales), post mortems revealed that a number of the oiled scoters (taken to the RSPCA hospital in Somerset) had died of haemorrhagic enteritis (Powell, personal communication). Sharp, too, notes that '[a]fter the Exxon Valdez oil spill, a significant proportion of oiled birds autopsied by the Madison Laboratory of the US Fish and Wildlife Service showed gastrointestinal haemorrhaging and pneumonia'. For further illustration of the difficulties, he used the (unpublished) data of D. M. Fry, who:

> reported that causes of death of 50 seabirds taken to rehabilitation centres after oil exposure in the Exxon Valdez spill included stress (due to oiling, cleaning, handling and captivity), haemolytic anaemia, emaciation, hypothermia and secondary infections in captivity (p. 226).

Two particular problems are encountered at the RSPCA centre in Somerset. Guillemots have a tendency to develop pressure sores on their toe joints and hock. And aspergillosis, a fungal disease, is common amongst the captive sea birds. The fungal spores proliferate in confined spaces and attack the birds' lungs.

The fourth problem peculiar to the rehabilitation of oiled seabirds is that treated and released birds are sometimes in danger of becoming re-oiled; even if released well away from the oil spill, they may make their way back there. After the Milford Haven disaster, RSPCA workers were concerned that treated birds, released off the south coast, would move north towards the oil spill. The decision was made, though, to release in the south because long-distance travel to a location north of the oil spill would be hazardous to the birds (Powell, personal communication).

Clearly, the rehabilitation of oiled birds (and particularly seabirds) is complex and fraught with difficulties and there is evidence to suggest that the survival rate is low. Sharp concludes in his paper 'Post-release survival of oiled, cleaned seabirds in North America' that this is the case. By using North American ringing data for marine birds, he tries to assess the effectiveness of the treatment and cleaning of oiled seabirds. His findings show that for oiled, cleaned and treated guillemots, the mean and median days survived were 39 and six days respectively – compared with mean and median days survived for non-oiled guillemots of 485 and 216 days respectively (p. 223). Survival from initial rescue to one year after release (based on the mean proportion of birds released combined with the data for post-release survival) was about five per cent (p. 226).

The problems I outlined above and the apparently low survival rate may lead one to suppose that the rehabilitation of oiled seabirds is futile and has no conservation value. Certainly, Sharp concludes that '[t]he available data ... do not support a decision that oiled seabirds should be rescued, treated and cleaned' (p. 227). However, such a conclusion is much too sweeping and too hastily made.

To begin with, we should recognise that a low survival rate of treated oiled birds is to be *expected.* It would be unreasonable to expect all – or even most – treated birds to survive on a par with birds that had undergone no such traumas. So to base justification of their rehabilitation on a comparison of their survival with that of non-oiled birds is unfair. The effectiveness and justification for proceeding with, say, heart by-pass operations are not based on patient survival as compared with the survival of *healthy* members of the population. One simply could not expect birds that had experienced oiling and all the attendant risks and traumas to survive as if none of this had occurred. Indeed, Sharp seems to be conflating two assessments: one of the effectiveness of the rehabilitation of oiled birds, and the other of the survival of oiled, rehabilitated birds compared with the survival of the population at large. To assess the effectiveness of rehabilitation, it would be fairer to compare the survival of oiled, rehabilitated birds with the survival of oiled *non*-rehabilitated birds. In a generous assessment one might say that one per cent of oiled birds would survive without intervention. If this were the case, then the five per cent survival rate for rehabilitated birds would represent an improvement of 500 per cent. This does somewhat change the perspective to one in which rehabilitation is viewed positively. And if the results of a similar study (Camphuysen *et al*, 1997) – in which survival rate was found to be 20 per cent – are considered, rehabilitation must be seen in a very positive light indeed. (Even in the worst case scenario, in which only one per cent of rehabilitated oiled guillemots were found to survive the first year (Wernham *et al*, quoted in Mead, 1997), rehabilitation could not be accused of doing any harm.) If, on the other hand, one is comparing the survival of oiled, rehabilitated birds with that of the population at large, then the oiled bird returns ought to be included in the 'population at large' returns figure. Again, this would show rehabilitation in an improved light.

In his paper, Sharp concentrates on the failures while mentioning the successes only briefly. He concedes that:

> there were a few recoveries suggestive of long-term survival of oiled birds after release i.e. a small number of oiled birds recovered after 6 months to 3 years, suggesting a limited dispersion of treated birds into the wild population (p. 225).

However, he immediately dismisses the importance of these finds in saying:

> [t]hese exceptions were too few to measurably influence mean and median days survived, and certainly the small number of birds involved would not contribute appreciably to the wild population of non-oiled birds.

Such a dismissal is misleading. As Mead (1997) reports, Harris and Wanless (1997) have found three rehabilitated oiled birds (two from the Netherlands and one from Germany) amongst the colony on the Isle of May in the Firth of Forth, Scotland. Yet it is not only that Sharp is too dismissive. He is also seen, here, to make an assumption, as he does not explain what he means by 'contribute appreciably'. Whether or not something is seen to contribute appreciably rather depends upon what ends are considered as worthwhile.

In quoting Chris Mead, the then Senior Ringing Officer with the British Trust for Ornithology as saying:

> [a]ll our experience with the rehabilitation of oiled birds points towards a very low success rate....the results from many hundreds (possibly several thousand) cleaned Auks – mostly Guillemots are really quite depressing ...

Sharp seeks to emphasise the uselessness of such rehabilitation. The hopefulness of Mead's claim that '"we do have a few ringing recoveries which show survival for several years and birds returning to distant potential breeding areas"' is not acknowledged. Once we see that the survival rates of de-oiled birds cannot be expected to compare favourably with the survival rates of unaffected birds, the actual survival rates can be viewed more positively.

So post-release survival, though significantly less than survival in the population at large, is positive and hopeful. And Sharp's study does not have universal application. As he himself points out, though there are 'surprisingly few data ... available' for assessing the effectiveness of the treatment and cleaning of oiled birds (p. 222), one of the studies 'supports the view that cleaning and treatment of oiled seabirds results in a significant level of rehabilitation' (p. 227). This was work with Jackass penguins Spheniscus demersus, 'a high percentage of which (37–84 per cent) were later seen alive at colonies (Morant et al, 1981)'. Whether these results indicated differences relating to species or to other factors such as type of oil, cleaning, or treatment methods is not known. What is clear is that the results from Sharp's work should not be awarded more significance than is appropriate for a single, rather limited study. Indeed, the RSPCA have been treating several swans in the Bristol dockland for oil contamination over three successive years, in each of which the swans have produced cygnets (Powell; personal communication). Mead similarly notes high success in the treatment of the mute swan, surmising that success may depend upon the type of pollutant.

From a conventional conservation perspective, the success with the Jackass penguins is especially significant given that they are an endangered species. The work quoted took place between 1970 and 1980. In 1994, Robin Maynard reported on South Africa's worst-ever oil spill, which affected Jackass penguins on Dassen and Robben Islands. At the time of writing, 'some 1,400 of the 7,000 penguins originally brought to the mainland for de-oiling made the journey back [and Maynard reports that] conservationists hope that 70 per cent of all treated birds will survive to be released back to the wild'. Indeed, Mead writes that successes with the penguins are high, 'with up to 84% of rehabilitated birds later being seen in their South African colonies'. It must be assumed that not only were the earlier oil spills learning experiences for the rehabilitators, but that the level of success in the 1970s allowed conservationists after the 1994 oil spill to be so hopeful.

It is easy to see how experience with the rehabilitation of oiled common species can benefit the rehabilitation of oiled endangered species and how even modest survival rates for the latter will be of conservation value. Also, many of the earlier arguments have emphasised the importance to conservation of engaged involvement with the natural world, the kind of involvement that rehabilitation offers. The rehabilitation of oiled birds is a particularly good example. An oil spill is a disaster by any standard but affects, most acutely, those living locally. Usually, however, there is little that can be done by local people to ameliorate the effects of the oil except the picking up of wildlife casualties – usually birds. Such rescue operations serve to assuage some of the frustration felt by those most touched by the disaster. Many people become involved in the rehabilitation effort, volunteering their services in catching, washing or caring for the casualties or in donating food, towels or cash to assist the work. When much else appears bleak, a rescue and rehabilitation effort engenders hope and provides an arena for positive action.

Our identification with the suffering of other sentient beings and our need for engagement and positive action are classically recognised by the media, who cover stories of suffering and rescue with equal relish. Sensationalist at times, perhaps, such coverage nevertheless serves to focus public (and government) attention upon the incident itself. It also raises awareness of the general issue of environmental pollution. While the bulk of environmental damage after an oil spill remains unseen and poorly understood, the death and suffering of birds is all to easily grasped, and illustrates the severity of the situation. The BBC One television series 'Back to the Wild' (August 1996) which followed rehabilitation work at the RSPCA Wildlife Hospital at West Hatch, Somerset, was an effective reminder of the damage caused by the oil spill at Milford Haven several months earlier. The individuals, whose mortality, suffering and rehabilitation are captured on film, are, in effect, ambassadors for the conservation cause. Appealing to emotions is an effective (and, given our make-up, legitimate and appropriate) way of raising awareness and educating the public in conservation matters.

Through focusing on the individual animals affected by a pollution incident, the effectiveness of the conservationists' wider remit – to conserve the natural world – is enhanced. The Co-operative Bank assumes that the plight of individual suffering animals can carry a strong environmental message, and that appeal to such suffering is a powerful medium through which to reach the public. At one time they ran an advert promoting their services as environmentally sensitive, using the close-up image of an oiled guillemot (*BBC Wildlife*, 1996, p. 13). Environmental groups such as the RSPB do not themselves care (in the practical sense) for the survivors of oil spills. However, their image – and the environmental message – benefits from close association with organisations, such as the RSPCA, who do. The RSPB states in the article 'Oil Outrage' (Anon, 1996a) that following the Sea Empress disaster, 'collaboration between the environmental organisations was excellent. As always, the RSPCA handled those oiled birds that were still alive, taking them to well-equipped specialist centres' (p. 16). A full-page photograph of a badly oiled common scoter (p. 17) accompanies the article, in which the RSPB vow to use the 80,000 'pledges of support for our marine conservation work ... to help convince the Government that our members want to see change'.

The rehabilitation of oiled birds offers us the chance of acquiring valuable knowledge about the way in which oil pollution affects wildlife. Observation, blood profiles, post-mortems – and the ringing of released birds – together with the collection and analysis of all such data – are important, not only for animal welfare but for conservation too. The RSPB admit that one of their aims is to '[f]ind out more about the long-term effects of oil on birds and their food supplies' (p. 16). There is no doubt that the process of rehabilitating birds will continue to add to this knowledge. Although Sharp concludes that rehabilitation is worthless from a conservation perspective, his paper could not have been written had rehabilitation not been practised. It is not only successes that convey valuable information; failures are equally informative. And of course rehabilitation practice produces both. To cease rehabilitation efforts would be to cut off an important and fruitful source of knowledge that will continue to have relevance to conservation practice. If there is one criticism to be made of rehabilitation, it is that too little follow-up takes place. Not all rehabilitators, for example, ring released birds, thereby missing out on potentially valuable information. It is only through knowledge of successes and failures that rehabilitation methods will improve.

Sharp claims that '[t]he available data ... do not support a decision that oiled seabirds should be rescued, treated and cleaned' (p. 227). He can see no conservation value in saving the small number of birds that his study determined are truly saved. Sharp's analysis of the situation is unacceptably narrow; he fails to recognise the full value of rehabilitation practice. The rehabilitation of oiled birds has, in addition to its value on humanitarian /animal welfare grounds, a distinctive and important value on conservation grounds.

The conservation value that rehabilitation engenders is not reliant upon procuring a high success rate. It could even be argued that, in terms of conservation value, a high success rate would be counterproductive. If the rehabilitated birds survived well, the effects of an oil spill would seem all too easily rectified. This could result in a more relaxed attitude from oil companies and the public alike. Admittedly, the majority of birds die from oil contamination prior to rescue operations, but attention is focused primarily on the live oiled birds and their progress. Given the scale of oil pollution in our seas, the most positive conservation measure must be prevention – tougher laws, higher fines, more effective policing of the seas, use of double-hulled ships – rather than more efficient clean-up operations. The fact that rehabilitated oiled birds may survive poorly is a salutary reminder to all that oil pollution is a serious environmental threat. If cleaned and treated oiled birds are not surviving well, then this reflects the devastation that must be taking place in the sea itself. Sharp makes a valid point in saying:

> Because their mortality rates are so high, oiled seabirds that have been cleaned and released should be added to the total of dead birds for the purpose of assessing damages of oil spills to seabirds and for the purpose of seeking compensation. Cleaning and treatment do not provide effective mitigation for and cannot be considered as even partial restoration of damage (p. 227).

That some birds *do* survive oil contamination engenders hope and necessitates the development of more efficient clean-up and treatment procedures. Yet the apparent poor survival (at least in some cases) alerts us to the serious nature of such contamination and maintains a focus on the need for preventive measures. The environmental cause is best served by a two-pronged attack of this kind.

Sharp's recommendation to cease the rehabilitation of oil pollution casualties is ill thought-out. While the conservationist's ultimate aim is to prevent oil pollution, the reality is that it is on the increase. What is disturbing is that while major oil spills receive mass media attention, oil is polluting our seas virtually all year round. The RSPCA Wildlife Hospital in Somerset deals with a steady stream of oiled birds from September through to April (Powell; personal communication). And the RSPB, (Anon, 1996b, p .5) writing about the Sea Empress oil spill, say:

> The disaster is unforgivable. But we know that routine, illegal discharges of oil from ships, month by month, continue to kill as many birds around our coasts, even if they rarely hit the headlines.

They go on to inform us that:

> At least three oil pollution incidents off the East Yorkshire and Humberside coasts between September and January affected more than 2,000 birds. The oil almost certainly came from ships dumping oily waste at sea.

It will be some time (if ever) before both major disasters and the illegal dumping of oil at sea are brought under control. In the meantime, our wildlife is exposed to a continuous – and increasing – barrage of oil pollution. To abandon rehabilitation efforts in such a climate seems nothing short of irresponsible. Our wildlife, exposed as it is to a range of modern pressures, needs all the help it can get.

In rehabilitation terms, the example of oiled birds can be considered a 'hard case'. This detailed study has nevertheless shown that, even in hard cases, my arguments for the compatibility of conservation and wildlife rehabilitation hold up under scrutiny. Given this, there is increasingly less reason to deny wildlife rehabilitation conservation status.

Conclusion

My claim that wildlife rehabilitation ought to be considered as a legitimate conservation strategy has not been made lightly. Following on from the arguments, in Chapter Seven, that wildlife rehabilitation has conservation value, I have dealt with the most serious worries of those who take a dissenting position. It turns out that wildlife rehabilitation is no threat to the conservation value of natural selection.

It turns out, too, that there is no need to worry about a conflict of interests with respect to conservation's familiar strategy of culling. Culling is neither required nor justified by conservation's holistic stance, the rationale most commonly dictating this practice. Other serious concerns about the conservation value of wildlife rehabilitation have been faced head on. The rehabilitation of oiled birds – perhaps regarded as the hardest case of all – has been critically examined and the criticism found wanting. My claim, then, that conservation and wildlife rehabilitation are both compatible and complementary remains a robust one. No evidence has arisen against my earlier, more fundamental claim that conservation ought to place greater emphasis on the value of the individual.

PART IV
THE UPSHOT; IMPLICATIONS FOR ACTION

Chapter 9

Managing wildness?

Introduction

In the preceding two chapters I have presented various arguments demonstrating the conservation value of wildlife rehabilitation. There are good reasons for claiming that conservation practice ought to take into account the welfare of individual wild animals in its policy decision-making and implementation of action. However, I argued in Chapter Five, that conservation's *raison d'etre* – the safeguarding of nature in something like its current, cherished, form – is best achieved by protecting wildness (where such protection allows for the *perpetuation* of wildness). Immediately, then, there is a tension. Whilst I have considered, and rejected, some important criticisms of wildlife rehabilitation, it could be objected that I have sidestepped the most fundamental criticism of all. The practice of wildlife rehabilitation may be seen to compromise the very thing both rehabilitation and conservation seek to protect: the animals' wildness.

In Chapter Five I demonstrated that our understanding of wildness is compatible with human proximity, activity or interference. Even though Europe, for example, has been extensively altered by human activity, many places within it are nevertheless considered to be wild. However, now I must ask a very specific question with respect to wildness and human activity. Namely: is it possible to reconcile the intensive management characteristic of rehabilitation practice with the conservationists' primary aim of protecting and promoting wildness? Can animals' wildness thrive under conditions that involve captivity and the close proximity of (and handling by) humans? In other words, does it make sense to talk about managing wildness?

Unravelling wildness

When we talk of wild animals we are commonly referring to their wildness in terms of genealogy. That is, if the animal is a member of a species that has not been domesticated, it is considered to be a wild animal. Consequently, the circus tiger and the zoo giraffe are considered wild in the same way as the lion living on the African plains. However, it is clear that for present purposes, this understanding of wildness is inadequate. Conservationists are not aiming to promote zoological parks (other than as a very short-term measure); the wildness they seek to safeguard involves much more than genealogical wildness. As became clear in Chapter Five, wildness entails not only wild animals but also systems of relationship within a particular story or narrative. The circus tiger and the zoo-living giraffe are in the wrong narrative. To put this simply, they are unable to *express* their wildness.

Though rehabilitation practice has quite different aims from those of circuses or zoos, it also shares some common features. The animals are confined to cages, away from their natural habitats and subjected to a degree of interaction with humans. Whilst mere proximity to humans does not necessarily compromise wildness, it could be objected that confinement and handling does. Whereas with circus and (most) zoo animals there is no end to this confinement, the aim of rehabilitation is to release the animals back into their natural habitat. Nevertheless, the nature of the rehabilitation process is such that it frequently causes wild animals to be tamed. An animal that has been tamed is denied the opportunity to express its wildness. It might be thought that the circus tiger's wildness is compromised – at least in part – by its having been tamed. Rehabilitation practice cannot be acceptable to conservationists if the methods and techniques used result in tame wild animals.

Wildness contrasted with tameness

On first consideration, this seems a reasonable objection, but on closer scrutiny it collapses. This is because of the confusion and misunderstanding surrounding the concept 'tame'. Tameness does not, as such, compromise wildness. We are mistaken if we imagine that an animal that flees from us is necessarily wilder than one that does not. Wildness is not synonymous with fearfulness and cannot be contrasted with tameness. An animal is simply as fearful – or as tame – as it needs to be, and this will depend upon the type of animal it is, as well as the particular circumstances.

When some of those animals that we might refer to as 'original' wild animals – those that we had never before encountered – first met humans, they showed no fear, only interest (see Clutton-Brock, 1987, p. 12; Darwin, 1905, p. 23; Turner, 1994). They had no need to fear humans, as they had not yet learned that we are predators. Their tameness was a perfectly natural wild behaviour. However, this might be rejected as a special case: our concern is with the degree of fearfulness or tameness of animals that have experience of humans. I noted earlier that numerous free-living wild animals (e.g. blue-tits and robins) exhibit a degree of tameness in that they allow humans to approach more closely than do some birds living away from human habitation (the nightjar, for instance). But this might be an example of begging the question, for there is nothing in this observation alone that proves these animals are fully wild. Their tameness may be an indication of their having lost a degree of wildness due to constant human proximity.

There are, however, examples of the non-urban kind that illustrate perfectly the relationship between tameness/fearfulness and survival need. Jordan (1991) says of skunks, for instance, that 'the first time you encounter a wild one, your first thought is that it's tame: it goes about its business without paying you the slightest attention'. But this behaviour is a result – due to its wonderfully smelly deterrent spray – of its supreme confidence (pp. 84–85). As Jordan observes '[i]n the light of its powers, its fearlessness makes perfect sense'. De Grazia (1996) argues that there are evolutionary reasons for believing that an animal's expression of – or absence of – fear (that is, its fear threshold) is a *survival* mechanism. Indeed, fear has a similar adaptive value to pain (p. 118). Where humans become a greater threat, wild animals' flight distance will increase to aid survival, such as was the case when humans developed better weapons for hunting. As Clutton-Brock (1987), discussing the history of domesticated animals puts it:

> ... animals learn quickly and it is only with modern sophisticated weapons that hunted animals will begin to flee from any great distance. All prey animals have a relationship with their predators so that, for example with the lions and antelope in Africa, as long as the prey can keep the predator warily in sight it will not try to escape. It is only when the lion comes too close that the antelope will flee (p. 14).

In other circumstances, survival is aided, not by the animal becoming more flighty, but by its being able to sustain close proximity to humans. For example, the red kite, recently reintroduced to parts of Britain, still carries a legacy of tameness from its adaptation, as a scavenger, in towns where it was once a common sight. In talking about the crows in and around the suburbs of South California, Jordan (1996) describes their 'balance between alertness and indifference'. He explains that the crow 'must be alert, but it must also be comfortable, it cannot tolerate the emotional trauma of an unceasing fight-or-flight response. Its temperament must therefore be adjusted toward a certain tolerance of human presence' (p. 87).

The familiar – and apparently relatively tame – house sparrow presents an interesting case. It may be thought that this species is less wild than, say, the sparrowhawk, because it tolerates human proximity. However, a meticulous study of this bird has revealed that it does not merely allow humans to come close while it goes about its business. Rather, its behaviour changes when it knows it is being watched; the sparrow becomes extra wary. As Denis Summers-Smith (1997) puts it:

> ... house sparrows are acutely aware of human presence. If you want to see them behaving naturally, the secret is to look at them from the corner of your eye. Switch to a direct stare and their behaviour will change.
>
> It's not that they're shy, it's just that they're very careful in their dealings with people ... This wariness is one of the key qualities that allow sparrows to live closer to man than any other wild bird (p. 36).

A similar phenomenon can be witnessed on the African plains, where many eyes within the herds of feeding ungulates keep watch for predators. It has been observed that often the herbivores will allow the predator to come very close, seemingly aware of when real danger lurks and when they are safe. To react to every false alarm would clearly be a waste of energy and a poor survival strategy. Closer to home is the example of the Scottish red deer that becomes tame in the winter when left alone and provided with food, but fearful in the summer when it is in danger of being shot (Wigan, 1993, pp. 132–3). It is clearly not the case that the different behaviour patterns exhibited by these animals reflects changes in their wild status. The degree of fear – or apparent tameness – exhibited by an animal is not a reliable indicator of its wildness.

In support of the claim that there is no necessary link between fearfulness and wildness, or between apparent tameness and domesticity, we need look no further than those domesticated animals that are deprived of socialisation with humans during a critical period for learning. Normally, within this time period, their capacity for multiple species identity allows them to direct their species-typical behaviours towards humans as well as members of their own species. If, however, the individual animal is denied access to humans during this critical period, socialisation will not take place (see, e.g. McCune *et al*, 1995). The animal will consequently remain fearful of humans throughout its life. This fear may be mitigated by habituation (positive exposure) to humans. But habituation is not equivalent to socialisation and so is not an alternative to it. Even those domestic animals that have had adequate socialisation with humans may become fearful of us if social contact is not maintained. A domestic animal, then, may be every bit as fearful of humans – perhaps even more so in some cases – as a similar wild animal. So the exhibiting of fearful/non-fearful behaviour is dependent upon firstly, the type of animal and how its survival is best promoted (is it an animal like the skunk with an effective deterrent or a vulnerable fast-running prey species?). Secondly, it is dependent upon the particular situation (is the herd of antelope faced with a hungry or merely inquisitive lion?). And lastly, individual experience is relevant. Have the individual's experiences been sufficiently broad to prepare it for the kind of life it is designed to live? Or are there glaring omissions in its past experiences that mean fearfulness is inevitable? For fear is exhibited – at least in part – as a result of events for which the animal's past experience has not prepared it. No generalisations can be made, then, about the relationship between fearful/non-fearful behaviour and the wildness (or otherwise) of an animal.

The non-fearful behaviour so far considered (seen, for example, in the house sparrow, but also in animals such as the urban fox) is what might be called pseudo-tameness.. It has come about as a natural behavioural change, part of the animal's own adaptive strategy. Where rehabilitation is responsible for behavioural changes of this nature – that is, where animals become less reactive to potential dangers – this, too, may reasonably be regarded as adaptive, *pseudo-tameness*. The animals have become habituated to humans; they have learned that, at least in their current environment, it is not worth reacting excessively to us. Indeed, it may be that this habituation, in reducing levels of stress, aids the rehabilitation process, allowing a more speedy return to the wild. This being the case, pseudo-tameness clearly promotes rather than compromises wildness.

If, however, we can make no clear distinction between wildness and reduced fear or pseudo-tameness, there is a distinction to be made between wildness and 'tameness proper', that is, the *tamed* condition. Indeed, a distinction between them must be found if the state of 'wildness' is to have any meaning at all. What, though, is the difference between these two states?

Two kinds of dependence

If what I have named 'pseudo-tameness' does not rob an animal of its wildness, then the state of being tamed surely does. Intuitively, it seems fair to argue that urban foxes and magpies remain fully wild despite living in close proximity to humans, whereas tamed circus tigers or domestic cats do not. The taming of an animal does seem to seriously compromise its wildness. The difference between these two states of tameness is, I suggest, one of *dependence*: the latter but not the former state is one of true dependence upon humans. It may be argued that it is possible to tame an animal without making it dependent upon humans, but I would reject this claim. It seems to me that part of the *definition* of 'tamed' is the notion of dependency upon humans.

If this is so, we face a problem. If a domestic animal is not wild because of its dependence upon us, what should be said about, for example, the house mouse? Apparently, it cannot survive without us; within a few years of humans vacating the island of St. Kilda, the house mouse became extinct (*Readers Digest Field Guide*, 1984, p. 139). Yet surely we would want to say that it is a wild species in the same way as the magpie and the blue-tit?

It may seem at first that it is necessary to argue that the house-mouse is not a wild species, for its dependence upon us is total, in-as-much-as it cannot make out in the wild without us. (This is in contrast with another species closely associated with us – the house sparrow. Since it nests occasionally 'in hedges and ivy-clad trees' (*Readers Digest Field Guide*, p. 274) it would presumably not die out if we did.) However, there is a significant distinction to be drawn between what we might call pseudo-dependency and true dependency. In the case of pseudo-dependency the species members have retained their natural behaviours and adaptations; with true dependency these have been replaced (at least in part) by human-dictated behaviours and adaptations. The reliance of the house mouse upon humans is of its own making; its dependence on humans is a contingent fact. For the house mouse, governed as it is by natural selection, the human is merely a feature of its environment to which it has become supremely and exclusively adapted. Indeed, there is no difference in kind between the reliance of the house mouse on its particular environment and the reliance of other wild creatures upon their environments. Each is equally tied to their environment for survival.

The difference in dependency between wild and tame animals is, therefore, not to do with either proximity to humans or the degree of reliance upon us. Rather, it is to do with *the way in which dependency has come about.* Where humans force dependence upon wild animals, they necessarily erode these individuals' behaviours and adaptations, replacing them with human-dictated behaviours and adaptations. Dogs, for example are modified by us (genetically and through conditioning and training) to comply with human needs and moral codes. Circus tigers are behaviourally modified according to a set of human standards. True dependence (as opposed to pseudo dependence) is deliberately instigated and perpetuated by humans. It is purposeful, being designed to benefit us. While the circus tiger, the domestic cat and the laboratory mouse are all, to varying degrees, subjected to such purposeful dependence, the house-mouse, though dependent upon us for food and shelter, is not; indeed, its presence in our houses is generally unwelcome. This is not to say that humans cannot influence the evolution of species like the house mouse; our use of poisons and traps has no doubt necessitated the evolution of adaptation for survival. However, we have not taken control of the house mouse and moulded it to fit with an end of our own choosing.

This distinction is not the same as that between natural and artificial selection. Circus tigers fall into the category of true dependency, yet they are not artificially selected for (except, perhaps, in a very marginal way). Indeed, dependency is applicable, not as selection is, to species, but rather, to the *members* of species. It does not make sense to talk of tame *species*, for even where domestication has taken place, there are examples of domesticated individuals, groups or populations that either never were tame or have left behind the tamed condition. As 'the original state', and that which contrasts with tameness, wildness is viewed in the opposite way. Species are considered wild by dint of their genealogy; the tiger is a wild animal whether living in an Indian forest or a British circus. However, as discussed earlier, there is more to wildness – as a state worth protecting – than mere genealogy. The wildness of species members, groups, or populations can be compromised. Exhibiting a greater tolerance of humans does not in itself count as compromised wildness, but the state of dependency – that which characterises true tameness – undoubtedly does.

According to genealogy, the house mouse is a wild species. There was a worry that, if dependency is the mark of tameness proper, either this species' wildness is in question, or dependency cannot, in itself, compromise wildness after all. We have found, however, that members of the house mouse species cannot be considered to be dependent upon humans in the true sense. The wildness of this species is therefore intact.

It could be objected that the question about the house mouse is misleading. Given that 'tameness' relates only to members of species while 'wildness' relates also to the species itself, there is a sense in which the wildness of a species can never be usurped by tameness. This is in fact true: the status of a species cannot be altered by the status of some of its members. Nevertheless, this does not mean the question about the house mouse is neither important nor worth asking. By addressing it, we have reached a better understanding of the criteria that determine true dependency. This in turn ensures that we are in a better position to identify instances in which the wildness of individuals has been compromised. As I have said, we are interested in wildness as it relates to both species (genealogy) and individuals. But further, the discussion raises an interesting and important question: what exactly, is responsible for the change in a species' wild status? If it had turned out that all members of the house mouse species were, in fact, tame, would this have been sufficient to alter the species' wild status? In other words, how many members of a species must be tame before the species is considered to be no longer wild?

Ordinarily, we consider domestication to be the factor that changes the status of a wild species. Characteristically, it is the norm in a domesticated species that its members fall into the category of 'dependency'. However, the defining feature of domestication is the alteration of species' characteristics through artificial selection. Yet although some domesticated species – such as the dog (but see Chapter Six) – have been highly altered through artificial selection, others, such as the reindeer, rabbit and cat have been altered remarkably little. It does seem, then, that the status of domestication is somewhat open to interpretation. It may, in some cases, constitute dependency at the individual level but not (genetic) manipulation at the species level. Since it could feasibly be argued that dependence upon humans is a sufficient condition for the removal of a species' wild status (domestication), it was clearly important to have demonstrated that members of the house-mouse species are *not*, in fact, truly dependent upon humans. To put this another way, since the definition of domestication sometimes rests upon 'dependency' rather than artificial selection, it was important to show that the dependency of the house mouse was not a sufficient condition for its being considered domesticated.

We have now arrived at a clearer understanding of how the wildness of individuals and species is compromised. With this in mind, I will go on to consider whether the practice of wildlife rehabilitation is guilty – even unwittingly – of such a compromise. However, before doing so, it is worth considering the status of an animal's wildness in a little more depth. What, exactly, is the *nature* of such wildness, and in particular, how fragile is it?

The nature of wildness

I have argued that we are mistaken in believing that an animal that flees from us is necessarily wilder than one that does not. An animal that increases its flight distance has not become *more* wild – it is simply *maintaining* its wildness; that is, its ability to cope in nature. It is merely taking account of a new situation in its environment (such as human predators with their ever-improving weapons). If we take any example of a wild animal – one that is currently living in the wild – we simply cannot *increase* its wildness. As Livingston (1994) puts it, 'wildness is *whole*' (p. 172). We can – and have – caused some wild animals to change their behaviour, but that is all. To say that we cannot make a wild animal more wild is not to say, however, that we could not make it more ferocious, more savage; certainly we might. But there is no simple correlation between wildness and ferocity. We might just as easily make a domestic animal more ferocious. Such an animal would, however, be no more wild than its docile conspecifics. This would simply be another example of a change in the animal's behaviour. In at least this one respect wildness is less amenable to human control than might at first have been supposed.

Some may object that in at least one case – the wolf – human impact has resulted in a wilder animal. It is speculated that when the wolf was first domesticated it had a wide template for socialisation in the sensitive period of its life (if similar to the dog, between approximately four and ten weeks of age). This meant that, along with other species that were successfully domesticated, it would readily socialise with humans. But because of our persecution of the wolf – our refusal to live alongside it – we have selected against such sociability in the wild wolf. It is believed that the modern wolf has a much narrower template for matching during the sensitive period than its ancestors that were described as having 'multiple species identity'. The fact that modern wolves may be less able to socialise with humans (see, for example, Zimen, 1987 and Woolpy & Ginsburg, 1967) might reasonably be thought to mean that the wolf is wilder than previously. This, however, would be another example of misunderstanding what it means to be wild. In narrowing the wolf's social template, we have, in fact, reduced its general adaptability – an important 'wild' feature. That is to say, we have reduced the wolf's ability to cope in the wild. Rather than increasing the wildness of the wolf, we may have decreased it. Even if one argues that the wolf is now better adapted to its hostile, human-populated environment by having reduced its former ability to socialise with us, the most one could say about this would be that the wolf has simply maintained its wildness. It still could not be said that we humans have increased the wildness of the wolf.

It we cannot make an animal more wild, we can certainly make one less wild. Does this mean that wildness is a fragile state, with, as it were, little defence in the face of human impact? Perhaps not, for it would seem to have a degree of resilience.

We can see this by considering the fact that, for example, a tame (as in tamed) wolf is not the same thing as a dog; it is not trainable and is extremely reactive.[1] Hearne (1994), for instance, writes that:

> ... no account I know of concerning work with wild animals gives useful advice for dealing with the fact that wolves, lions, tigers, orang-utans and chimpanzees remain willing to commit mayhem no matter how large their vocabularies (p. 23).

Wildness can, to a degree, withstand human impact. It has a certain robustness.

Even where wildness has been fully compromised (as in domestication), it retains the ability to bounce back. Although a large proportion of individual domestic animals may be unlikely to survive in the wild, most (if not all) species of domestic animal have members that have succeeded in becoming feral (living wild). Jordan (1991) informs us, for instance, of the surprising fact that '[a] survey done in the San Francisco Bay area found that about 60 percent of unaltered [un-neutered] house-cats become feral within three years' (p. 99). Dogs, too, can become feral remarkably easily. In Italy, populations of feral dogs rely upon an influx of dogs from nearby villages (Boitani *et al*, 1995). From personal experience of working with rescue animals, dogs abandoned by their owners can quickly revert to the feral condition. One interesting case concerned a very docile pet Shetland Sheepdog that inexplicably turned feral for a period of time until eventually coaxed back to its former state by the owner (Masson, 1997).

Some authors mistakenly over-emphasise the difference between wild and domestic species. Livingston (1994), for instance, believes that domesticated animals have lost the ability of their wild cousins to form inter and intra-specific relationships. He refers to the 'experiential undernutrition' of domestic animals, meaning by this, the animals' 'failure or inability to assimilate to any significant degree the experience of self-as-group, while self-identity as either multispecies community or biosphere is beyond it altogether'. He says further, that such an animal 'cannot mature as a fully sentient and participatory being' (p. 119). It is probably true that domestication has brought about changes in animals' perceptual worlds (or as Hemmer originally called it, animal's *Merkwelt*). That is, whereas a high degree of alertness or sensitivity 'combined with quick reactions to stress are essential for the survival of an animal in the wild, the opposite characteristics of docility, lack of fear and tolerance of stress are the requirements for domestication' (Clutton-Brock, 1995). Nevertheless, it is not plausible to suggest that domesticated animals lack the ability to form rich social relationships; it is more likely that they rarely have the opportunity to develop and express them. Consider, for instance, two pieces of evidence that illustrate the very powerful intra and inter-species bonding of domestic cows. The first concerns a two-year-old Devon cross Friesian heifer. Having been sold separately from her calf, and the two housed on farms seven miles apart, she was found the next morning suckling her calf, having negotiated unfamiliar territory during the night (Valentine, 1997). The second concerns a

---

[1] Reactivity is an important survival mechanism for wild animals. They must be able to respond instantly to dangerous stimuli.

remarkable farming system in the Cotswolds, where two hundred cows are allowed to live in a way that farmed cattle never are. They are virtually free-range, often live into their twenties, and form relationships with whichever cows they choose. What has been allowed to develop is 'a system of matriarchal family groups'. In one instance, after a difficult birth in which the calf was stillborn, the cow sought out her own mother for comfort, even though 'the two had not associated with each other for a year or more'. In another instance, one of the cows woke the farmer (Rosamund Young) in the night and led her to her seriously ill calf two fields away. As Nicolson (1995) reports:

> [t]hese two stories illustrate the cow's ability to find the essential help it needs. It shows that there is a habit of inter-reliance among cattle, a bonding of mother and daughter that persists through long periods of being apart; it also shows that the Youngs are part of that network, bound into the social system of their cattle (p. 1).

The wild and domesticated states, then, are not so different as to prevent wildness re-emerging from the domesticated condition. At the same time, wildness is sufficiently robust to prevent its easy erosion, such as through the action of taming.

It may be objected that in order for a docile, domesticated animal to regain its wildness, it must experience fear of humans. This problematises the earlier claim that wildness is not synonymous with fearfulness. I would not wish to deny that the development of such fear is sometimes a factor in releasing a tamed animal from its dependent condition. For example, a frightening experience (e.g. abandonment) may be the catalyst for a domesticated animal's moving to the feral state from which it continues to be wary of humans and their association with the feared experience. However, in other cases, fear is not the driving force. The feral state may be entered gradually, where dependence on humans is slowly replaced by independence, such as may occur when house cats become feral. Even if it were argued that in any one individual, fear of humans must be a factor in the transitional stage from the tamed condition to the wild or feral condition, it would still not follow that fear of humans makes the difference between the wild and the non-wild state. It remains the case that some wild individuals are less fearful of humans than some domesticated individuals.

An idea is beginning to emerge of the kind of thing an animal's wildness is. It is neither a static condition nor as prone to manipulation and compromise as might have been supposed. However, if I am right in supposing that the perpetuation of wildness is the means by which the conservationist best protects the natural world, I need to be more specific about what, exactly, wildness is. How, then, can it be best characterised? What, in other words, is the nature of wildness as it relates to animals?

One suggestion might be that wildness is a *character trait* or *disposition*, in the sense that one may be described as having 'a sunny disposition' or 'a mean streak'. In that they describe tendencies within an individual, these categories are inadequate to describe 'wildness'. An individual may have a sunny disposition but be quite capable of moving freely in and out of this condition. Though wildness is not necessarily a static condition it has greater stability than aspects of an individual's character. It cannot be displaced in the way that dispositions so readily are. Indeed, it is something over and above an individual's set of dispositions. Wildness relates to something more about an individual than a single facet of their character.

Another possibility for capturing the nature of wildness is that of a behaviour. Certainly, it is commonly said that an animal is 'exhibiting wild behaviour'. Too often, though, this seems to be used when an animal is exhibiting fearful or aggressive behaviour towards humans. This aside, whilst 'behaviour' may be closer than 'disposition' to capturing the sense in which 'wildness' relates to the individual as a whole, it too, fails to take account of the degree to which wildness is immune to change. An animal's behaviour may instantly change (e.g. from play to predatory behaviour) in a way that an animal's wildness cannot. Unlike behaviours, an animal's wildness does not change rapidly in response to different needs.

Perhaps the most successful characterisation of 'wildness' is its description as a state (as in, for example, a state of poverty). This certainly captures the idea that, within certain parameters, wildness is constant, while at the same time leaving open the possibility of change. This might be further articulated as a 'state of being'. Wildness, then, is without doubt, something *of* the animal. It is a form, if you like, of self-expression. Nevertheless, a circus tiger is not wild in the sense that its conspecific in the Sumatran forest is. The difference between the two is that the latter, but not the former, has the means by which to express its wildness to the full. And it is this *expression of wildness*, this 'making out' as a wild animal that conservationists and protectors of the wild need to preserve and perpetuate. While the wildness of animals (in the formal, genealogical sense of wildness) can be perpetuated in a zoo, this offers us a pale shadow of wildness as a meaningful state – that is, wildness in its fullest expression. This is the sentiment of Lemmon's (1993) article about wild chimpanzees, a sentiment captured in its title 'A chimp in a cage is nothing'. In particular, part of a wild animal's self-expression is its interaction with the natural world, where it takes its place in the ecosystem. As Cooper (1993) eloquently puts it:

> ... what we admire (envy, perhaps) is ... the wild animal's complete possession of its environment. Its harmony with its surroundings is no mere adaption to them – through camouflage, say, or ingenious thermostatic mechanisms. It consists, rather, in that unreflective ease of movement and activity, that practical knowledge "in the hands" (or paws or claws) which, in a different modulation, is also what we admire in the traditional craftsman (p. 342).

A wild animal's self-expression entails, too, almost by definition, the exhibiting of its full behavioural repertoire. It also entails the opportunity to reproduce – that is, the opportunity for the perpetuation of its own kind.

We have reached, then, a fuller understanding, not only of the ways in which wildness may be compromised, but also of the nature of wildness. Though without doubt vulnerable to human impact, wildness has about it, a certain robustness or resilience. There may even be said to be a pull towards wildness. The tamed condition is not self-perpetuating in the way that the wild condition is, for it requires continuous human input, there being a tendency towards the state of ferality. As a condition that spans both fragility and resilience, wildness is best characterised as a state of being, a state of being that can find expression to different degrees depending on the environmental conditions. Having reached this fuller understanding we are now in a position to consider the practice of wildlife rehabilitation and its relationship to wildness. The question to answer is this: is wildlife rehabilitation guilty of compromising the wildness of animals?

Wildlife rehabilitation and the perpetuation of wildness

I argued, in Chapter Five, that human impact, as such, is not considered to be incompatible with wildness. However, as discussed in this chapter, certain kinds of human impact can and do have an adverse effect upon wildness. Though wildness has a degree of resilience, where an animal's dependence upon humans is deliberately sought and maintained, its wildness will be compromised. When we consider that the practice of wildlife rehabilitation requires just such deliberate (and sometimes prolonged) dependence upon humans, it would seem reasonable to fear that it may compromise an animal's wildness. More worrying still, while some forms of human impact have a temporary effect upon an animal's wildness, other forms have, what is thought to be a *permanent* effect on wildness. One such form of human impact, relevant to wildlife rehabilitation, is of particular risk, and this is *imprinting*. Imprinting is the process by which an animal comes to recognise humans rather than others of its own kind as kin. Since imprinting is a potentially serious threat both to the success of wildlife rehabilitation as well as to the acceptance of wildlife rehabilitation by conservationists, its causes and effects must be investigated and fully understood. It is to this task that we now turn.

Imprinting

Wildlife rehabilitation almost always requires incarceration of the wild animal and its increased exposure to humans. It is arguably true that incarceration, in itself – even with proximity to humans – does not seriously compromise wildness. I acknowledge that captivity often changes behaviour. Even a fish – the sergeant major fish (*Abudefduf saxatilis*) – behaves slightly differently when in captivity. This is thought to be due to the effects of a controlled closed system (Prappas *et al*, 1991). Similarly, in captive-raised mustelids, though there is a rudimentary capacity to kill prey (Manning & Dawkins, 1992, p. 40), skills necessary for locating prey may be eroded during captivity. Captivity often closes off opportunities for behaviours. Nevertheless, it could be argued that captivity has little long-term impact upon animal behaviour. For example, there is much evidence that skills can be re-learned (Barton, 1994a). As one zoo director, talking of condors says, condors have been evolving for millions of years and so it is unlikely that a few generations in captivity would change them. One convincing example involves the case of migrating birds. For example, the grey heron and Branta branta – both of which escaped from Stockholm zoo – though kept in cages for several generations, migrated normally on release (Hallander, 1976). The black-footed ferret provides another example. After being reared for several generations (five years) in captivity, at least five ferrets – reintroduced to Shirley Basin in south-eastern Wyoming – survived their first winter. At least one female went on to breed (Barton, 1992, p. 63). Clearly, rehabilitation does not involve inter-generational captivity. It involves captivity in relation, only, to parts of individuals' life spans. Captivity and the attendant acceptance of close proximity to humans (or pseudo-tameness) are not, in themselves sufficient to compromise wildness. However, the potential for imprinting is.

The phenomenon of imprinting is not fully understood and hence 'cannot be precisely defined' (Manning & Dawkins, p. 35). It might best be described as a learned response that occurs very early in life enabling young animals to recognise and become attached to a parent figure and/or a future mating partner. Or, Lieberman (1993) describes it in very broad terms as 'the process by which some species learn to recognize other members of their species' during an early period of their life (p. 392). It is unlike other learned responses in that it does not rely upon physiological rewards such as food or warmth. Imprinting plays a part, then, in primary socialisation (Sluckin, p. 53) and sexual choice (p. 9). Though originally imprinting was a phenomenon narrowly associated with birds (and particularly species such as geese, ducks, and chickens), it is now recognised as having application to the young of various mammalian species (Sluckin, p. 13), and perhaps even all mammals and birds. From the fact that imprinting occurs, we might conclude that at least some species of animals do not instinctively recognise their own kind. As Sluckin notes, Lorenz (1935) observed that, when reared in isolation, the young of many birds 'did not respond later in a normal manner to members of their own species; and when reared by human beings, they behaved towards people as if people were their natural companions' (pp. 49–50). Incorrect imprinting is clearly a potential danger where young/orphaned animals are raised in captivity. However, since it has proved possible to raise very young animals and successfully rehabilitate them to the wild, there is no reason to believe that inappropriate and lasting imprinting is inevitable. Indeed, it is possible to identify those factors that predispose an animal to such imprinting. By so doing, it is possible to suggest ways in which unwanted imprinting might be avoided. It is also possible to offer tentative explanations for why some instances of wildlife rehabilitation are more successful than others. It is worth looking at these factors in some detail because the need to avoid imprinting in rehabilitation practice is critical. Imprinted animals cannot be released into the wild.

Initial effectiveness of figure It does seem that some species, at least, have an inherited predisposition to respond towards their own species, but this innate bias can readily be overridden (Manning & Dawkins, p. 39). It is important for rehabilitators to learn about the species they are rearing and to avoid exposing youngsters to stimuli that will encourage attraction/attachment to humans. Although moving objects are more attractive, attachment to stationary figures has been recorded (Sluckin, p. 65), and noisy figures are more effective than quiet figures (p. 28). Rehabilitators ought, then, to consider the viability of rearing young animals without their being exposed to humans. This will clearly be possible in some cases (e.g. ducklings, since they eat unaided) but not others (e.g. un-weaned orphan fox cubs that must be hand-fed).

Exposure time Laboratory-based experiments have revealed that the depth of imprinting is related to the degree of exposure to stimulation (Sluckin, p. 52). It may, then, be possible to avoid significant imprinting despite hand-feeding youngsters, if contact time is kept to a minimum. However, the more quickly contact can be withdrawn (both on each occasion and ultimately) the better.

Exclusiveness of exposure It has been found that in communally-living ducklings, a strong following response waned much earlier (ten days) compared with singletons (three weeks). Similarly, chicks reared in groups were found to imprint less well on objects than single chicks (Sluckin, pp. 51–52). That is, there is an important social element in imprinting behaviour. Certainly, from personal experience, singly-reared birds do seem particularly prone to becoming lastingly imprinted on humans, and there is some evidence to suggest that badger cubs reared singly are more likely to imprint on humans than those reared with other badgers (Powell; personal communication). This strongly suggests that wherever possible, rehabilitators ought to rear young animals with others of their own kind.

Although social rearing of youngsters is to be recommended, it is pertinent that in the laboratory, it was found that ducklings followed (and became imprinted on) objects more readily if in the company of other ducklings. It would therefore be inadvisable to allow the mixing of individuals inappropriately imprinted upon humans (or perhaps, any other object) with normally behaving individuals. This social tendency might, however, be used to advantage to encourage normal imprinting in certain individuals. That is, it may be worthwhile mixing, say, a single inappropriately imprinted youngster with several normally imprinted conspecifics.

Critical period Imprinting takes place during what is known as the critical period, with the strongest imprinting of all occurring at the peak of this period. For example, it has been shown that puppies form lasting relationships with people or dogs only if exposed to these attachment figures between the ages of three and seven weeks (Sluckin, p. 70). Sluckin writes:

> Scott (1962) holds that critical periods are of the utmost importance in determining the social, emotional and even intellectual development of any individual ... early learning determines later behaviour and inhibits any subsequent re-patterning of behaviour.

While at one time this period was believed to indeed be truly critical (for example, Sluckin identifies Hinde, 1962), some authors now claim that the period for imprinting is never sharply defined, being optimal rather than critical and thus better described as the 'sensitive period'. It is not, Sluckin says,

> ... a period that is circumscribed solely by physical maturation; for its range is determined by the impact of the environment upon the developing organism. For example, the critical period will be extended in those individuals that have not, as yet, had the chance to become imprinted. That is, if no stimulus is present, the animal will remain responsive for longer (p. 78).

For example, as Manning and Dawkins (1992) note, '[s]ocial rearing restricts the sensitive period and it may be that the imprinting process itself plays a part in ending this period' (p. 37). Nevertheless, of the different learning processes, 'the most critical learning period of all has appeared to be that for imprinting' (Sluckin, p. 71). It is therefore well worth rehabilitators learning about critical/sensitive periods (and their peak times) for the species they deal with and taking extra precautions during this period of an animal's life. Critical periods can differ quite markedly, even in apparently similar species. It is also important to understand that keeping youngsters in a barren environment during the critical period for that species, in order to prevent inappropriate imprinting, might simply postpone the problem.

It is also worth noting that there are some indications that imprinting could occur in older precocial birds. Sluckin cites some experiments where social learning occurred in chicks at later developmental stages than normally associated with imprinting and concludes that '[t]he possibility of imprinting playing some part in this later social learning could not be excluded' (p. 81). Nevertheless, there is, as yet, no conclusive evidence 'that in precocial birds the capacity for imprinting can persist, even under the most favourable circumstances, beyond the juvenile stage of development'. With mammals, however, it may be a different story. Sluckin mentions Harlow's (1958) reports that 'filial attachments of young monkeys to surrogate 'mothers', while normally acquired very early in life, could also develop [at a diminished intensity] at the comparatively late age of eight months' (p. 81). The circumstances of these laboratory experiments (where the young monkeys, prior to exposure at eight months to cloth mothers, were denied alternative stimuli) are unlikely to have relevance to real life and to rehabilitation. Perhaps, too, the other instances of possible imprinting in adult animals have little relevance to rehabilitation, 'the best-attested such occurrence [being] in she-goats and ewes which become attached to their young soon after parturition' (p. 112). Nevertheless, the rehabilitator should be aware that imprinting may not solely be an issue that relates to a single, very early critical/sensitive period.

Individuality As with all forms of learning, with imprinting there can be substantial differences not only between species responses but between individual responses, even when individuals have similar 'histories' (Sluckin, p. 31). Rehabilitators ought not, then, to treat their charges in too 'clinical' a fashion. They should always be aware of the possibility that individuals will not conform to expected patterns of response or development.

Multiple species identity Whether exposure to humans during an animal's sensitive period leads to inappropriate imprinting may depend in part upon whether or not that animal is a member of a species with 'multiple species identity'. That is, whether it has the ability to form attachment to more than one species. Puppies, for example, can become strongly attached to both dogs and humans if exposed to both from an early age. And Sluckin writes that '[t]here is evidence that young birds can become attached to more than one figure at once' (p. 54). For those individuals with 'multiple species identity' it may not be sufficient, as a means of preventing inappropriate imprinting, to provide them with conspecifics if they are also exposed to humans. Individuals that do not possess 'multiple species identity' on the other hand, may fare better, but only if provided with a suitable alternative attachment figure (e.g. a sibling). Without this provision, such individuals will presumably be at great risk of becoming imprinted on humans, and may do so even if exposure to people is very minimal. The implication is that animals with 'multiple species identity' could be at greater risk of imprinting on humans than animals that do not possess multiple species identity. However, it might also be speculated that, provided they also form appropriate attachments, their chances of wild survival are much greater than those of an individual that fails to make any appropriate attachments. In other words, if after dual imprinting takes place the opportunity for perpetuating the attachment to humans is subsequently denied, attachment to the appropriate figure will flourish and eventually eclipse the attachment to people.

It is possible that this is, at least in part, what happens in cases where rehabilitators, who form strong bonds with their animals, nevertheless succeed in returning them to the wild. Ratcliffe (1983), for example, had close relationships with the badgers she reared; yet they eventually led entirely wild (independent) lives. However, it may be that another factor is relevant here: some rehabilitators who bond closely with their animals ensure that the bond is an *exclusive* one. That is, they do not allow their animals to imprint upon the human *species* but only upon themselves; no other human contact (of a positive kind) is allowed. This is certainly what happened in the case of rare bald or waldrapp, ibis (*Geronticus eremita*). Their reintroduction to the wild turned from failure into success when the scientists came 'to appreciate the importance of the tight social and family bonds in this colonial bird' (Barton 1994 b, p. 12). The scientists took over the role of parent ibises such that:

> The chicks were fed every two hours, always by the same two researchers, with insects, chicks and mice. At 20 days they were taken to a specially adapted farmhouse, where they had a view of the surrounding fields and mountains. Here they were cared for by their foster parents, one of whom was with them at all times, as is normal in parent birds.

The crucial point is that the ibises were looked after, exclusively, by the two scientists. Once, Barton reports, 'Dr. Pegoraro was amused to see all six ibises fly to a stranger who looked like her from behind. As soon as the woman turned round, though, the birds realised their mistake and flew off'.

It could be argued that an upbringing that provides such close bonds allows the animal to develop more closely to the way it would have if it had experienced maternal bonds. And provided this bond is allowed to diminish just as the maternal bond would naturally have done, the animal may be better adjusted than one deprived of a close bond. As Manning and Dawkins (1992) observe, '[i]mprinting-like phenomena are clearly involved in the social development of mammals' (p. 36). Orphan bats, for instance, very quickly deteriorate if denied human company. Bonding, as opposed to generalised taming may, then, facilitate rather than impede successful rehabilitation.

Sexual imprinting The phenomenon of imprinting is associated, not only with attachment but also with later sexual choice. There is disagreement about when sexual imprinting occurs. Sluckin claims that it occurs earlier than attachment (p. 9), while Manning and Dawkins believe it to occur later than filial imprinting (p. 39). Inappropriate sexual imprinting could be problematic for successful rehabilitation. Sluckin writes: 'Lorenz (1955) stated that birds of many species, when imprinted to man, would refuse to copulate with "conspecifics"'(p. 56). And according to Lieberman (1993), Lorenz 'imprinted a male jackdaw on him when it was young, and found that when it reached sexual maturity it attempted to seduce him by offering him worms that it had thoughtfully predigested' (p. 368). However, Sluckin observes that 'such fixations are not often exclusive', noting the example of a female domestic pigeon that, though reacting sexually to a man, 'was later successfully paired to a male domestic pigeon'. The female bird did, however, continue to react to the man even after acquiring a mate of its own kind. So it seems that once sexual attachment has been established it may never be entirely forgotten.

However, not all cases of sexual imprinting follow this pattern. Lieberman writes about Guiton (1966) who 'found that chicks that were exposed to a moving yellow glove when young later attempted to copulate with the glove in preference to a stuffed female chicken'. But he goes on to say that 'if they were then allowed to live with other chickens rather than remaining in isolation ... they developed a preference for the chicken model'. Clearly, says Lieberman, 'the preferences established during early life can be modified by later experience' (p. 370). So whilst inappropriate sexual imprinting may prove to be problematic for the rehabilitation of a wild animal – even if later appropriate sexual attachment takes place – there is a chance that it will not. But more importantly, it seems unlikely that sexual attachment to humans could develop to any degree in the first place if normal precautions against inappropriate filial attachment are observed.

Reinforced conditioning Imprinting is a unique kind of learning in that it is not a response to physiological rewards such as food or warmth; learning takes place solely as a result of response to a stimulus. However, rehabilitators ought to be aware that subsequent conditioning can reinforce attachment. If, for example, a young animal that had begun to imprint on humans were, in addition, to associate humans with food, its attachment would be strengthened. It seems likely that reinforced conditioning is responsible for the enduring attachments that are formed between some animals and humans.

Clearly, where young animals require hand feeding, it is not possible to avoid a human/food association developing. Indeed, it is vital that hand-fed orphan animals come to trust the rehabilitator quickly in order for them to feed adequately and to avoid stress. However, it is imperative that the link between humans and food is broken as early as possible if successful rehabilitation is to be achieved. Where it is not, and the animal learns to depend upon humans for resources, the bond may become irreversible. As wildlife rehabilitator Stocker (1989) notes:

> Most patients, even some with horrific injuries, which are brought in to us, make a perfectly good recovery. The one affliction for which we have no cure is that of being tame. Whenever the cardboard box discloses a tame bird or animal, we know from the outset that it could never survive in the wild and can never be released (p. 154).

Well-meaning members of the public frequently rear orphaned animals. However, they often remain unaware of the need to encourage independence in their charges as early as possible. The animals consequently develop a strong association between humans and food. Nevertheless, there is ample evidence that the mere formation of an association between humans and food need not prevent successful rehabilitation: many thousands of hand-reared animals of a variety of species are rehabilitated annually. The problem arises when the association is fostered for too long.

Medium inhabited The degree to which an animal becomes imprinted upon humans could feasibly be partly determined by the medium it inhabits, or even just the habitat it ultimately frequents. The ease with which dolphins – even those that have been in captivity for many years – are rehabilitated may be due, in part, to their inhabiting a world quite separate from the human world. Johnson (1990) documents the successful rehabilitation, in 1987, of two Atlantic bottle-nosed dolphins, Joe and Rosie, by the Oceanic Research and Communication Alliance (ORCA) in America. He writes that 'their successful return to the wild proved even more remarkable since the animals had already been languishing in captivity for seven years' (p. 8). More recently, in Britain, the dolphins Rocky (Morecambe Marineland), and Missie and Silver (Yorkshire's Flamingo Land) were successfully returned to the wild after many years of performing in captivity. In the case of Joe and Rosie, a certain amount of 'de-training' took place before release. Nevertheless, it could be that because dolphins inhabit such a radically different world to us, this acts as some kind of natural barrier that inhibits, or minimises, strong imprinting. Alternatively it may allow for a more rapid readjustment to the wild. Pryor (1995) highlighted the difficulty of training captive dolphins by any other means than through a hands-off system of positive reinforcement. We are quite incapable of forcing a dolphin to act in a certain way, because we are aliens, so to speak, in their watery world.

In a similar way, it could be argued that one of the reasons orang-utans are more successfully rehabilitated than chimpanzees (see later) is that they inhabit thick impenetrable forest in which they infrequently come into contact with humans. The process of losing their attachments to humans is consequently relatively straightforward. In contrast, chimpanzees are generally thought to have very poor rehabilitation prospects, and it could be that this is in part due to the fact that their wild habitat overlaps with the human habitat. Attachments are more difficult to break when proximity is maintained.

Reversibility At one time, imprinting was thought by some to be an irreversible process. For example, Sluckin quotes Lorenz:

> it is too early to claim definitely that the imprinting process is irreversible. I infer that it is so from a fact frequently observed in hand-raised birds. Once their instinctive social reactions are transposed to a human being, their behaviour does not change in the least even if they are kept for years with other members of their own species and without human company (p. 8).

However, laboratory studies indicate that this is not necessarily the case. For example it was shown that 'socially reared chicks, subsequently kept in isolation, often respond all the more vigorously to *any* moving object' (Sluckin, p.40). This illustrates that '[i]n one sense, therefore, their initial imprinting was highly reversible' (p. 80). Similarly, Livingston (1994) makes the point (considered above) that 'unless [imprinting] is systematically reinforced, the attachment wears off shortly' (p. 124).

Yet even in cases where animals born in captivity form an attachment to humans that is strengthened by positive reinforcement (such as feeding), all may not be lost. This can be seen to be true even of the wolf, a species that has multiple-species identity. And, notwithstanding the earlier point that we may have reduced wolves' sociability, they may be seen as having a special affinity with us. Writing of the North American red wolf reintroduction programme, De Blieu (1993) says that, on release, '[t]he animals did not seem to understand that trucks and cars could harm them. More discouraging, a few of them were not at all frightened of direct contact with people' (p. 69). Nevertheless, the wolves did, in time, seem to lose this familiarity, so much so that when one female roamed outside the designated reserve and had to be caught up and returned, the biologists did not find it easy, and would never have succeeded without the help of telemetry. The biologists, says De Blieu, did not anticipate 'how cunningly wild she would behave'. Indeed, three months after the release, the wolves 'were seldom spotted from the roads'. And not only were they becoming more secretive, they were also proving more effective hunters in that they 'began to show evidence of being able to kill prey larger than rabbits and racoons' (pp. 76–77).

The reversibility thesis seems the more plausible given the above discussion of factors that help to determine the strength of imprinting. Degrees of strength imply degrees of durability. Indeed, Sluckin suggests that though irreversibility is no doubt usual under 'natural' conditions, it is, 'relatively rarely attained otherwise. Imprinting appears to vary in extent or thoroughness like other forms of learning' (p. 57).

The degree to which imprinting is reversible must be dependent not only upon the situation in which the individual finds itself but also upon the natural developmental pattern of the species to which it belongs. For example, although the chimpanzee and the orang-utan are relatively similar species, their rehabilitation successes differ markedly. As mentioned above, the chimpanzee is said to be extremely difficult to rehabilitate (McGreal, 1993), while with the orang-utan, many successes have been demonstrated (Borner, 1979). This may not only be to do with habitat differences but also with bonding and social development. It is not that the young orang-utans bond with their human carers less than young chimpanzees do. Indeed, young orang-utans are totally dependent upon humans. The important difference may be to do with detachment from the primary carer. In the orang-utan – a largely solitary animal in the wild – detachment is a naturally evolved behaviour. Once weaned from its primary carer, the chimpanzee, on the other hand, continues to live in a close-knit society for the duration of its life (Goodall, 1986). Certainly, some of the species most prone to irreversible imprinting belong to the very sociable corvid family (e.g. crows, magpies and rooks.) Stocker (1989), for example, writes about a hopelessly imprinted magpie that was never successfully integrated with the wild population despite being given its freedom (pp. 156–159). Other species of bird, though reared in a similar way to the corvids, will reach a point in their development when they begin to pull away from the rehabilitator, refusing to take food from them and eating on their own (personal experience). And one species, at least, must have an in-built resistance to one form of imprinting: the cuckoo, though reared by an individual of a species other than its own, does not become attached to or sexually imprinted upon this species. There may, feasibly, be all sorts of species-specific subtleties that affect the strength and reversibility (or otherwise) of imprinting.

One other factor affecting reversibility of imprinting could be connected to whether the species is long or short-lived. As Manning and Dawkins explain:

> With short-lived animals, which must grow rapidly and are sexually mature at a few months of age, behavioural development is highly compressed and events like imprinting come to take on an all-or-nothing, irreversible appearance ... But the infancy of some mammals is very prolonged and development is far more a continuous process ... Experience has effects at all stages, some transient, some permanent, but rarely appearing as irreversible because there is usually time for later experiences to change the direction of development.

This example illustrates the point that the reversibility – or otherwise – of imprinting is not always a straightforwardly factual question, but sometimes a *relative* question, the answer to which is a matter of where we draw boundaries.

Although inappropriate attachment to humans is considered the primary danger for rehabilitation, other forms of imprinting are relevant too. Sluckin notes that '[v]ery little is known about the influence of imprinting upon the permanent preferences of animals for one or another kind of home environment' (p. 120). He quotes the example of Hopfer, who:

> found that wild [chipping sparrow] adults preferred pine-tree to oak-tree sites. So did birds reared without any foliage, pine or oak. Those that had been reared in the presence of oak foliage chose almost equally between pine and oak perch sites.

Sluckin wonders 'to what extent other species' preferences for habitats in adult life may be influenced by their infantile experiences' and surmises that perhaps 'early social experiences are more crucial for later social behaviour in some species than others'. There is clearly a need for the rehabilitator not only to prevent the inappropriate attachment of rehabilitees to humans, but also to foster appropriate environment-based attachments. While in the example above, the chipping sparrow's preference for oak or pine foliage may not be of vital importance for their survival, other preferences might, feasibly, make the difference between survival and non-survival. For example, it is possible that flocking behaviour (an important anti-predator strategy) may be a result of imprinting (Sluckin, p. 67). The more the captive habitat resembles a species' wild habitat the better.

Taking on board the factors that influence imprinting, it is possible to suggest some recommendations for the prevention or minimisation of inappropriate attachment during the rehabilitation process. The rehabilitator could usefully:

- Keep human contact – visual and audible – to a minimum.
- Provide conspecifics whenever practicable.
- Provide habitat and foodstuffs as close as possible to those encountered in the wild.
- Make use of foster parents if available.
- Consider wild-based rearing methods (e.g. feeding fox cubs in their earth).
- Avoid mixing naive youngsters with those imprinted on humans.
- Gain knowledge of the species dealt with including their imprinting behaviours.
- Avoid or minimise strengthening any inappropriate imprinting through positive reinforcement.
- Release at the earliest opportunity.

These are general precautions. Nevertheless, they are not always possible or practicable to implement and may sometimes be counterproductive. For example, as mentioned earlier, some rehabilitators deliberately form close and specific attachments to particular animals. Such bonds may facilitate successful rehabilitation, perhaps by fulfilling the animal's psychological needs, thereby allowing normal development. But it is important for rehabilitators to appreciate species-specific (and individual) differences: this particular rearing method may suit some animals and not others. It has been used successfully with badgers, but from personal experience it does not appear to be suitable for some other species, such as pigeons. Though the chances of inappropriate attachment will be minimised by following the above recommendations, it may yet occur. It is difficult to identify any hard and fast rules in the area of rehabilitation.

Conclusion

To recap, I have suggested that there are four forms of human interference that compromise animals' wildness, these being:

1. Domestication (replacing animals' adaptations genetically);
2. Taming proper (replacing animals' behavioural adaptations);
3. Permanent captivity (depriving animals of the opportunity to express wildness);
4. Imprinting (inappropriate learned attachment to a human (or humans in general) rather than the parent figure).

The process of rehabilitation has the potential to compromise animals' wildness but does not do so of necessity. There is a danger that rehabilitation may lead to attachment to humans through imprinting, thereby creating problems for the animal's successful reintegration with the wild population. However, once the process of imprinting is fully understood, the risk can be greatly minimised. And even in those cases where inappropriate imprinting has occurred, there may still be a possibility (provided the imprinting has not become too established) that this process could be reversed. Rehabilitation, in other words, does not necessitate forced dependency (tameness proper), the state characteristic of, say, domestication.

It is not, then, at odds with the conservationists' primary aim: the protection and promotion – or the safeguarding – of wildness.

Paradoxically, though, there may be a case to argue for the claim that some of conservation's very own management strategies compromise, or come close to compromising wildness. It might be wondered, for instance, whether the management regime operative on the Scottish Natural Heritage (SNH) National Nature Reserve of Rum is guilty of this. Here, the red deer research programme seems little inspired by, or directed towards, a respect for the animals' capacity for self-expression. The deer are variously segregated and restricted, and differential culling controls their breeding and group sex ratios. The deer are denied the opportunity to express their wildness to the full. Most significantly of all, perhaps, is the fact that the deer research is motivated, not by any obvious conservation end but by a desire to discover the most efficient management practice that will produce large-bodied, large-antlered stags for shooting. In other words, the purpose of the deer research is to determine how best to make the deer more manageable. So not only is the wildness of the deer compromised during the process of the research, there is no longer-term objective (as there is with wildlife rehabilitation) to rectify this situation.

Specific conservation management strategies may be guilty, then, of degrading wildness. So too might certain projects within a pivotal area of conservation: wildlife reintroduction. On the face of it, wildlife reintroduction would seem to be a paradigm case of the promotion of wildness. Yet despite its lofty ideal (wildness regained or reinvented), at least sometimes, the process of wildlife reintroduction is such that the ideal either cannot be attained or is severely tainted.

To illustrate, two examples will suffice. The first is that of the Californian condor. In 1989, the last of the three remaining condors was captured and 'for the first time in two million years there were no free condors in California' (Williams, 1992). Then, in January 1992, two eight-month old birds were released in the Los Padres National Forest, 75 miles north of Los Angeles, 'marking the end of a five-year absence from the Southern Californian skies'. So far so good. However, as the author notes, there is little in the way of *risk* in the new lives of the birds. The two are:

> fitted with radio transmitters and are continually monitored [such that] biologists will always have a good idea of where they are and what they're doing.
>
> And so that the vultures don't succumb to the factors that killed off their forebears – lead poisoning from the bullets in abandoned game carcasses, poisoned predator bait, high-voltage power lines and raptor-shooting ranchers – they will be fed for the rest of their lives on stillborn calves at the same spot in the middle of their sanctuary.

The author notes that 'reintroductions will continue until there are at least two separate self-sustaining populations of about 100 birds each'. However, it is not clear whether the close monitoring and feeding regimes will continue indefinitely or whether they will cease at this point.

If human control subsequently ceases then it could be argued that this merely represents an extended version of the necessary (but temporary) control required in wildlife rehabilitation. This is a state of affairs I have defended as both legitimate and ultimately respectful of wildness. However, it could also be argued that since such control – unlike wildlife rehabilitation – would be *inter-generational*, it goes beyond the bounds of acceptable human impact. Inter-generational human impact may be viewed as unacceptable because of its similarity to the kind of impact imposed upon domesticated animals. That is, human impact would not only affect individual animals, it would also affect the species and its evolutionary development. This aside, if, once the condor were self-sustaining, human control continued, there would certainly be a case for arguing that this constituted a compromise of wildness. There may not be a significant difference between the condor's managed existence and its caged existence. Indeed, the definition of a cage could feasibly encompass the former.

The second example that serves to illustrate the limitations of wildlife reintroduction as a perpetuator of wildness is that of another North American species, the red wolf. Finally, after many unsuccessful efforts to save this species in situ, the remaining wild-living individuals were caught up for a captive-breeding and reintroduction programme (De Blieu, 1993). Those red wolves that made it back to the wild were, however, subjected to a catalogue of human controls. They were, at various times: vaccinated against parvovirus, supplemented with food, medicated against heartworm and hookworm, treated when ill, fitted with radio-collars containing tranquillising darts, implanted with both radioactive chips and with hormones to prevent them from breeding. Their every movement was charted, and when they strayed outside the designated boundary they were unceremoniously bundled back. Nor did it end there. In order to maintain genetic diversity, individual wolves would be moved in and out of the area for some time to come; in other words, the wolves' breeding partners were human-dictated. These red wolves had become perilously close to living as caged animals. They had become perilously close, too, to living, in a sense, as domesticated animals – that is, animals for whom artificial and not natural selection operates.

My intention is not to argue conclusively that some conservation management – that includes wildlife reintroduction – compromises wildness. Rather, it is to point out the potential for hypocrisy. Should conservationists be reluctant to accept the arguments expressed in this chapter – that wildlife rehabilitation need not compromise wildness – they will at least be obliged to accept that it is no worse, and probably a good deal better, than many of their own management strategies in respect of safeguarding wildness. This said, the worries identified here relating to wildlife management suggest the need for a re-appraisal of certain of their strategies. If it is true that 'protection of wildness' is the conservationists' central tenet, then there may well be a need to alter, even abandon, some forms of wildlife management.

Chapter 10

The new conservation

Introduction

The main thrust of my argument throughout the book has been that conservation requires a new direction. It requires a different emphasis; one, primarily, that acknowledges the conservation value of the individual animal. I have arrived at this position in part by criticising some of the existing conservation mores, and in part by demonstrating that concern for the individual both precedes and feeds into the kind of concern that is needed for conservation action. This latter perspective has been supported by an examination of the practice of wildlife rehabilitation. This is a practice that might by seen to epitomise concern for individual welfare, and one that is traditionally regarded as having little or no relevance to conservation.

But this traditional assumption is mistaken. I have shown that there are subtle but unmistakable underlying links between conservation and concern for the individual, as well as very explicit practical links between conservation practice and wildlife rehabilitation. If there is one clear implication from these arguments, it is that conservation ought to support the practice of wildlife rehabilitation. This may mean that wildlife rehabilitation becomes part of the conservation agenda, that is, that conservation offers active support for wildlife rehabilitation. Or it may simply mean that conservationists cease to criticise the practice of wildlife rehabilitation, thereby offering passive support. I will return to the theme of wildlife rehabilitation shortly. But first, I will consider what other implications for conservation have arisen from the various arguments presented. I will consider, too, the ways in which the arguments support one another and lend further credibility to my central argument that wildlife rehabilitation has conservation value.

Resumé

I have assumed, throughout, that the purpose of conservation is the safeguarding of nature in a form that sufficiently resembles its current, cherished, form. Chapter One left us in no doubt that positive action of some kind is, in fact, required. The nature of the current extinction event is such that lost species are no longer being replaced. Were we to do nothing to remedy this, nature as we know and value it may well be lost. However, Chapter One also illustrated that the key to safeguarding nature is to concentrate, not upon the safeguarding of the species themselves, but upon the safeguarding of the opportunity for speciation. What is of real importance is the perpetuation of the variety or diversity. In Chapter Five, I argued that this aim is most usefully translated into the more concrete goal of protecting and promoting wildness. That is, where wildness is able to flourish, there exists the opportunity for speciation to take place.

Though some conservationists argue that nature is best protected by withdrawing from it and leaving it undisturbed, in Chapter Four I suggested that the only realistic way of achieving the conservationists' aim of safeguarding nature is through the actions of the 'thinking interventionist'. With respect to wildlife rehabilitation, then – clearly a care-giving interventionist practice – no immediate problem arises in its being supported by conservation. The potential worry that the interventionist position might not be compatible with the more specific conservation aim of safeguarding wildness was laid to rest in Chapter Five. There is no necessary clash between human involvement and the perpetuation of wildness. Nevertheless, it seemed that there might be cause for concern that wildness would be compromised by the kind of human involvement required of wildlife rehabilitation. But this worry, too, was dismissed, this time in Chapter Nine. Happily, provided rehabilitation is carried out in a sensitive, informed manner, wild animals can maintain their wild status.

My claim that the opportunity for speciation and not the protection of existing species ought to be the primary goal of conservation practice was given further credibility in Chapter Six. Species were found to be neither real units in nature nor important units of evolution. Rather, the species is a useful tool with which to categorise, explain, and study the natural world. Given this, I argued that species ought only to be referred to in a descriptive – and not prescriptive – sense. Conversely, the individual is a category that is both real and meaningful to us and so ought not to be prioritised by the category of 'species'. Here again, then, the potential resistance that the individualistic practice of wildlife rehabilitation may meet from the holistically driven conservation practice is found wanting. Indeed, conservation's holistic stance is heavily criticised in Chapters Six and Eight, demonstrating that those 'wholes' with which conservation is currently so concerned (primarily species) ought not to be prioritised in the way they are. Conservation's holistic categories cannot be truly delineated. Nor are there good conservation reasons for deciding whether one form of holistic category – such as forest – is better than another, such as moorland. The main implication of this for conservation is that culling can no longer be justified unquestioningly. Culling is based upon the assumption that certain 'wholes' are more valuable than others, either in themselves (e.g. species over subspecies) or in their particular form (e.g. forest habitat over moorland habitat). In order for culling to be sanctioned, conservationists would need to show why it is that an arbitrary grouping (e.g. a species), or a particular form of that grouping (e.g. a forest habitat), are more important to conservation than that which would prevail without culling. Conservationists would also have to show why these groupings are more important than the lives of individuals. By problematising the conservation practice of culling, a potentially serious barrier to conservation's acceptance of wildlife rehabilitation fell away. This was further reinforced by the demonstration that wildlife rehabilitation does not, as some seem to suppose, compromise evolutionary fitness.

If it is the case that the category of 'species' is no longer of such importance to conservation, it follows that the central place rare and native species currently enjoy must be reassessed. Irrespective of this point, the categories of 'rare' and 'native' were heavily criticised in Chapters Two and Three respectively. In Chapter Two, there were found to be good reasons for paying more attention to rare species where they are threatened, because a diversity of species is essential as 'stock' from which speciation is to occur. But it was argued that where rare species are not threatened, it is important to realise that these species are no more valuable than other such categories such as 'common species'. There are two implications that arise from this for conservation. The first is that conservation is forced to acknowledge the value of the common, something that it too seldom does at present. The work of conservation groups such as the charity Common Ground ought to be given greater priority. They are 'working to encourage new ways of looking at the world to excite people into remembering the richness of the common-place and the value of the everyday ... ' (Common Ground).

The second implication arising from Chapter Two is that there is greater reason still to argue that conservationists ought to accept wildlife rehabilitation as a legitimate conservation strategy. Recalling Chapter Seven, wildlife rehabilitation is non-discriminatory; it deals with the members of any species. Where these are the members of threatened rare species, such help can be vital to their species' survival. However, the vast majority of species with which wildlife rehabilitation deals are not, in fact, rare. Yet once conservation acknowledges the different but equal value of common species, wildlife rehabilitation practice takes on new conservation meaning and worth.

Just as rarity is problematised, so too, in Chapter Three, is nativity. Here I found that there are reasons to proactively protect species that can be said to belong in an ecological community. But the boundaries between (what are currently understood to be) native species and alien species are indistinct, and so there are no good reasons to actively discriminate against (that is, to harm) the (sentient) members of so-called alien species. Killing requires strong justification and no such justification is possible where the category such killing is designed to protect is an arbitrary one. This is so because, as discussed in Chapter Six, we are constituted to respond to and care about other individuals.

Once again, there are implications for the relationship between conservation and wildlife rehabilitation. Just as a large proportion of the wild animals dealt with by rehabilitators are members of common species, so too are many of them members of introduced, rather than native species. Currently, this is one source of tension between conservation and wildlife rehabilitation. Conservationists have argued that rehabilitating members of introduced species is of no conservation value because introduced species are themselves of little or no conservation worth. With our new understanding of the categories 'native' and 'introduced', this tension is dissolved. There is no longer any need for conservationists to disapprove of wildlife rehabilitation on the grounds that much of its work is with members of introduced species.

Through critical examination of a number of conservation's current values and strategies (species and other 'wholes', rarity, nativity and non-intervention) new ways of thinking about conservation, and new sets of emphases within conservation, have emerged. Though never the main theme of discussion, the notion of biodiversity has always been kept in sight, and has remained of central importance to conservation. This said, it has become clear that, because various notions associated with biodiversity (e.g. 'species' and 'rarity') have been criticised (and new ways of considering them proposed), the notion of biodiversity itself has to be differently understood.

For example, because the concept of 'species' (and its associated concept of 'nativity') has emerged as somewhat arbitrary in nature, biodiversity's emphasis should shift away from individual substantiations of diversity and concentrate, instead, upon the means of maintaining diversity in nature – that is, speciation. But it is not merely that the notion of biodiversity requires a shift in emphasis. It requires, too, the incorporation of a completely new idea: that individuals are important units of conservation. So while speciation is the means by which biodiversity is maintained (and as such, is biodiversity's mainspring) the individual, as the unit upon which relationship is based, is the mainstay of biodiversity. That is, relationship is required in order for those natural processes that maintain and generate biodiversity (adaptation/evolution/speciation) to operate. And, as we have seen, the individual is the unit that so often sparks off our care, passion and action in the conservation arena. In the absence of an opportunity to help actual individuals, the call to protect biodiversity would be much more likely to fall on deaf ears. Biodiversity remains a central conservation value, but its meaning has shifted, broadened and been enhanced.

The upshot

Conservation thinking, then, requires something of an overhaul. The 'new look' conservation might be characterised in the following way:

- Conservation is a human-centred activity with human-centred aims, objectives and decision-making procedures. Given that there are no clear conservation-based reasons for determining whether one species, or one habitat is better than another, a more relaxed attitude to human-caused ecological change is adopted.
- The boundaries of the categories 'native' and 'species' (and indeed holistic categories in general) are arbitrarily drawn, and will command less importance than currently is the case. The appellation 'rare' is not suggestive of particular conservation value. This means focusing more positively than is currently the case upon 'the common' (as is done, for example, by the conservation group Common Ground). It also means ceasing to sanction the harming (or even downgrading) of 'the introduced'.
- Acknowledgement of the arbitrary nature of the category 'species' will lead conservationists to shift emphasis away from *instances* of diversity (i.e. species) and towards the *means of maintaining and promoting diversity* – speciation. This will mean that the safeguarding of species is replaced in importance by the safeguarding of habitats – areas within which speciation can take place.
- Primarily because of its requirement for large areas of habitat in which nature predominates (the one most vital ingredient for speciation to take place), wildness will be perceived as the means of safeguarding nature, thereby

replacing biodiversity as the conservationist's main goal. The concept of biodiversity is difficult to articulate or pin down because it rests upon arbitrarily defined holistic notions such as 'species' and 'habitat'. Although maintained as an important conservation idea, it will be demoted from its central place as conservation's primary goal. Conservation policy will also take on board the notion that wildness and human activity are not necessarily mutually exclusive, and dispense with any notion that the only way to save nature is through the exclusion of humans from areas of supposed wilderness.

- By becoming aware of the arbitrary nature of holistic categories such as 'species', conservationists will become more aware of the importance of the individual as a real and meaningful unit. This will require that they, for example, re-think their policy on culling. Culling could not be sanctioned solely on the basis of attempting to save another species. Indeed it is doubtful whether the conditions that might be argued to legitimise culling for the sake of conservation (where a species was causing the extinction of all local – perhaps island-based – variety) would ever, in fact, by met. Certainly, though, the routine culling that currently takes place in the name of conservation could no longer be condoned. Not only is the individual a unit that is meaningful to us as the kind of creature we have evolved to be, it is important, too, as the core unit for *relationship*. In this capacity it is invaluable to conservation as inspiration for care, passion, and ultimately, conservation action.

In the process of questioning the current conservation rationale, the need for a new conservation has arisen. This is one that makes room for the importance of the individual, and one, therefore, that ought to take the practice of wildlife rehabilitation seriously. Wildlife rehabilitation, we have to conclude, has conservation value. However, implicit in this claim is the understanding that successful wildlife rehabilitation has conservation value. My final task, then, must be to determine what, exactly, successful wildlife rehabilitation amounts to. It is crucial that rehabilitation does, in fact, succeed. It is crucial, too, that rehabilitation's requirement for success accords with conservation's ultimate aim – and presumed requirement for success – helping to safeguard nature.

Determining successful wildlife rehabilitation

One possibility for determining successful wildlife rehabilitation is at the point of release. Some rehabilitators do seem to regard this as marking the point at which their work is done. But this is mistaken. Unless released animals have retained or learned skills for wild *survival* they may quickly perish. For the wildlife rehabilitator to achieve his or her aim of helping the individual, it is important to have some indication of the individual's survival chances. This entails putting into practice the kinds of procedural recommendations discussed in Chapter Nine, both during captivity as well as during the release process. Release itself is not a legitimate determinant of successful rehabilitation.

If the rehabilitator is to truly help the disadvantaged wild animal, the criterion for success must clearly be *survival* in the wild. That is, it is important for rehabilitators to know whether the animal they have released has the potential to survive for some length of time. It would not, of course, be possible to follow the fortunes of every individual. Instead, post-release survival studies ought to be carried out. Groups of individuals – representative of either species or of rehabilitation procedures – should be followed and monitored. Such studies have been undertaken with, for example, guillemots (e.g. Sharp, 1995), hedgehogs (e.g. Morris and Warwick, 1994) and foxes (e.g. Robertson and Harris, 1991). Some rehabilitation centres (for example, the RSPCA Wildlife Hospitals) attach British Trust for Ornithology (BTO) rings to their birds prior to release. Nevertheless, insufficient studies have been conducted. This is, perhaps, the most serious of wildlife rehabilitation's weaknesses. However, one area in which the question of post-release survival is receiving the attention it deserves is with an RSPCA study of rehabilitated (de-oiled) guillemots. Following the papers (such as Sharp's) that have demonstrated a low survival rate for such guillemots, the RSPCA decided to spend around £100,000 on a study to determine why survival is so poor. This involved using sea pools designed for salmon farming on which to study the treated birds, and analysis of their captive diet (Powell, personal communication). Results from the completed study will help to determine a future direction for oiled bird rehabilitation.

Such studies are vital if we are to understand what is necessary for survival, because sometimes we are mistaken in assuming which factors are most important for survival. It was found, for example, that black-footed ferret survival[1] was more dependent upon the inability to hide from predators than the more obvious ability to hunt for food. The latter skill was quickly learned whereas the former was not (Barton, 1994). Similarly, a study of rehabilitated hedgehogs revealed that, contrary to expectation, suburban gardens and not woodland provided a better chance for survival (Burch, 1995). And in another study, released hedgehogs were found to suffer from gum disease, pointing to the need for dietary changes while in captivity (Coughlan, 1996).

It may seem obvious that for rehabilitation to be regarded as successful, the individual concerned must not only survive whilst in captivity but must also be able to survive in the wild. However, it is not so obvious what is to count as wild survival. One understanding of 'wild survival' is the production of offspring. There are numerous examples of rehabilitated animals successfully breeding. For example, a rehabilitated group of badgers produced three cubs in each of 1994 and 1995 (Whitcher, 1995), and similar breeding successes in rehabilitated badgers have been recorded by the RSPCA (Powell, personal communication). As another example, Schmidt (1997), discussing the rehabilitation of oiled penguins in South Africa writes:

> The South Africans have documented sightings of penguins that were cleaned and ringed 22 years ago. Tony Williams, of Cape Nature Conservation ... says: 'Not only are we

---

1 This study was part of a reintroduction programme, but the survival issues are the same as for rehabilitation.

rehabilitating the birds, but we have evidence that a substantial number are entering the breeding population ... ' (p. 43).

The ease with which the species can be monitored subsequent to release clearly has a bearing upon the availability of breeding data. The swan is relatively easily monitored, and so is a good source of such information. In a study of the effects upon the mute swan of oiling followed by rehabilitation (Collins et al, 1994), it was found that, in 1988, two breeding seasons following the oil spill, 'there was no significant difference between oiled and control groups in the proportion breeding ... or in breeding success ... '. Nor was there any 'significant difference in the production of young during the subsequent two-year period 1989–90 ... ' (p. 87). If we are to count the production of offspring as the criterion for successful rehabilitation, it is clear that rehabilitation not only can be – but also frequently is – successful. This is so, particularly when we consider that a great deal more rehabilitated animals than is known about are likely to survive to breed, since follow-ups are both difficult and far too few.

There can be no doubt that the production of offspring is a sufficient condition for successful wild survival, but it is not clear that it is also a *necessary* condition. Not all members of a wild population are destined to breed. For some species, breeding is restricted (even for those of breeding age) to certain individuals within the population. This is the case with, for example, foxes. Factors such as food availability and population decline help to dictate the proportion of breeding vixen (Macdonald, 1989; p. 93). In other cases, breeding occurs but is unsuccessful. Non-breeding individuals are nevertheless vital members of the population. They may assist in the raising of others' young, but even when they do not, they are fulfilling a role in that wild community merely by existing in it. They are serving as competitors with which other potential breeding individuals (of their own species as well as others) have to contend – if not for mates then certainly for food. In this way they fulfil an important role in the evolutionary process. By surviving in the wild they are either eating or being eaten, thereby playing a part as interacting members of their ecological community. The production of offspring cannot, then, be viewed as the only indicator of successful wild survival.

Perhaps the only reliable indicator of successful wild survival is survival itself. Yet how long must an individual have to survive, once released, in order for the rehabilitation to be considered a success? Paradoxically, survival *per se* may not actually be very relevant. Consider, for example, the rehabilitated hedgehogs, discovered (through radio tracking) to have been eaten by badgers so soon after release (Warwick, 1993). The author did not regard them, necessarily, as failures. He notes:

> ... the fact that both wild and newly released hedgehogs fall prey to badgers suggests that it was not lack of experience that was their downfall. Badgers and traffic are a danger to all hedgehogs, wherever they come from (p. 55).

If an individual's actual survival is not always relevant to determining successful rehabilitation, then the statistical probability of survival is – that is, the expected chances of rehabilitated animals' survival. Yet what ought the survival chances of rehabilitated animals to be compared with? In Chapter Eight, I raised the point that it is inappropriate to compare the survival of debilitated, rehabilitated animals with the survival of non-debilitated members of the wild population, because the latter have not been subject to the illness/trauma/stress/isolation from the wild that the former group has. It is not unreasonable to expect rehabilitated animals to survive less well than non-debilitated/rehabilitated animals. The only appropriate comparison, then, is that between the survival of debilitated rehabilitated individuals and debilitated non-rehabilitated individuals. If rehabilitation improves the survival chances of such animals, it can be regarded as having been successful. Successful rehabilitation, then, makes a difference to the expected survival of debilitated animals. Although no studies have been carried out to ascertain to what extent rehabilitation increases debilitated animals' survival chances, common sense dictates that such animals' chances will, in most cases, be improved. Rehabilitation is, after all, a practice that largely rectifies our mistakes, that offsets the human impact.[2] It is a practice that serves to *remedy* a situation.

Conclusion

We have seen, already, that wildlife rehabilitation has conservation value in the sense that it provides an essential link between our innate sense of caring and the natural world. Conservation depends for its wider human involvement on the notion of care; maintaining this link is vital. The evolutionary roots of caring are arguably in the mother-infant relationship. One route by which we come to care about (and for) the environment is through caring about (and for) individuals that depend upon the environment. We have also seen that wildlife rehabilitation has conservation value in that – in a practical sense – it helps conservation to fulfil certain of its goals. This may be so in a direct way, such as when rehabilitation makes a difference to the survival chances of a threatened species. Or it may be so indirectly, when, for example, focusing upon rehabilitation helps to secure a habitat. However, we can also now say that the aims of wildlife rehabilitation and the aims of conservation dovetail nicely together. Successful wildlife rehabilitation, which is clearly the aim of all rehabilitation practice, is to increase the chances of debilitated animals' wild survival. This complements conservation's ultimate aim – that of safeguarding nature. In aiming to safeguard nature, the new conservation seeks to secure the opportunity for speciation through safeguarding wildness. The wild animal, as we saw in Chapter Five, is a key ingredient, as it were, of the wildness conservationists seek to secure. If the wild animal's chance of survival is enhanced, this can only serve to protect and promote wildness. Given that the process of speciation does not necessitate the breeding of every individual in a population, there is no need for the conservationist to demand a more stringent requirement of success from wildlife rehabilitation – such as, for example, the production of offspring. Wildlife rehabilitation and conservation are two distinct practices. However, given the shift

---

[2] Indeed, one source, (Steinhart, 1990) records that rehabilitators believe 80–90 per cent of their animals and birds come to them as a result of human activity.

in focus that I have argued is now required of conservation, it is clear that wildlife rehabilitation and conservation are a great deal more closely allied than has, until now, been supposed. My hope is that the two practices work ever more closely together, sharing not only their experience and expertise, but many of their values too.

Bibliography

Adams, W. M. (1996), *Future Nature: A Vision for Conservation*, Earthscan, London.

Aitken, Gill (1998), 'Extinction', *Biology and Philosophy*, Vol. 13, No. 3. July.

Anon (1994), 'Orkney Seal Rescue', *Care for the Wild News*, No. 7, Summer.

Anon (1995a), 'In Brief', *New Scientist*, 148, No. 2004, p. 13.

Anon (1995b), 'League slams ruddy duck cull', *Wildlife Guardian*, 31, p. 11.

Anon (1996a), 'Oil Outrage', *Birds*, Vol. 16, No. 2, Summer, pp. 16–17.

Anon (1996b), 'Oil Kills Wildlife', *Birds*, Vol. 16, No. 2, Summer, pp. 16–17.

Banks, Tony (1996), 'The Natural History Program', *BBC Radio 4*, 19 May.

Barn Owl Trust (1995), Information leaflet No. 6, Waterleat, Ashburton, Devon.

Barton, Miles (1992), 'Black-footed ferret rides again', *BBC Wildlife*, Vol. 10, No. 9.

Barton, Miles (1994a), 'Digging for Victory', *BBC Wildlife*, Vol. 12, No. 2.

Barton, Miles (1994b), 'Fostering the Future', *BBC Wildlife*, Vol. 12, No. 1.

Bateson, Patrick (1997), 'The Behavioural and Physiological Effects of Culling Red Deer', Executive Summary of Report to the Council of the National Trust.

BBC 1 Television (1996), 'Back to the Wild', August.

BBC Wildlife, (1996), Vol. 14, No. 7.

Begon, M. Harper, J. L. and Townsend, C. R. (1990), *Ecology, Individuals, Populations and Communities*, Blackwell Scientific Publications, Cambridge, Massachusetts.

Birch, Thomas H. (1990), 'The Incarceration of Wildness: Wilderness Areas as Prisons' *Environmental Ethics*, Vol. 12, Spring, pp. 3–26.

Blundell, John (1998), 'Letters' *BBC Wildlife*, Vol. 16, No. 7.

Boitani, L. (ed) (1976), *Proceedings of the Seminar, Reintroductions: Techniques and Ethics*, World Wildlife Fund, Rome.

Boitani, L., Francisci, F., Cuicci, P. and Andreoli, G. (1995), 'Population biology and ecology of feral dogs in central Italy', in J. Serpell, (ed), *The Domestic Dog; its evolution, behaviour and interactions with people*, Cambridge University Press, Cambridge, pp. 217–244.

Borner, Monica (1979), *Orang Utan; Orphans of the Forest*, WH Allen & Co. Ltd, London.

Bowler, P. J. (1992), *The Fontana History of the Environmental Sciences*, Fontana Press, London.

Brennan, Andrew (1988), *Thinking About Nature; An Investigation of Nature, Value and Ecology*, Routledge, London.

Brown, Paul (1995), 'Feathers fly over calls to cull alien birds' *The Guardian*, July 1, p. 8.

Budiansky, Stephen (1995), *Nature's Keepers: The New Science of Nature Management*, Weidenfeld & Nicolson, London.
Burch, Kevin (1995), 'Hogwatch', Wildlife Symposium, Lower Moss Wood, Cheshire, 5 August.
Cady, Michael and Hume, Rob (eds) (1992; 1988), *The Complete Book of British Birds*, The Automobile Association, Basingstoke, Hampshire and The Royal Society for the Protection of Birds, Sandy, Bedfordshire.
Callicott, J. Baird (1986), 'On the Intrinsic Value of Nonhuman Species', in Bryan Norton (ed), *The Preservation of Species; The Value of Biological Diversity*, Princeton University Press, Princeton, New Jersey.
Callicott, J. Baird (1989), *In Defense of the Land Ethic*, State University of New York, Albany.
Callicott, J. Baird (1995), 'The Value of Ecosystem Health', *Environmental Values* 4, pp. 345–361.
Campbell, N. A. (1987), *Biology* (second edition), Benjamin/Cummings, California.
Camphuysen, C. I. Duiven, P. Harris, M. P. and Leopold, M. F. (1997), *Sula*, 11, pp. 157–174.
Cater, Bill (1990), 'The Orang Gang', *BBC Wildlife*, Vol. 8, No. 8.
Caughley, G. and Sinclair, A. (1994), *Wildlife Ecology and Management*, Blackwell Scientific Publications, Boston.
Chalmers, A. F. (1978), *What is This Thing Called Science?* Open University Press, Milton Keynes.
Clarke, Stephen R. L. (1994), 'Is humanity a natural kind?' in T. Ingold, (ed.) *What is an Animal?* Routledge, London.
Clutton-Brock, Juliet (1987), *A Natural History of Domesticated Mammals*, Cambridge University Press, British Museum, London.
Coghlan, Andy (1996), 'Injured hedgehogs killed by human kindness', *New Scientist*, 3 February, p. 10.
Collins, Richard, Brazier, Hugh, and Whelan, John (1994), 'Rehabilitating a herd of oiled mute swans *cygnus olor*', *Biology and Environment: Proceedings of the Royal Irish Academy*, Vol. 94 B, No. 1, pp. 83–89.
Common Ground (promotional leaflet), Seven Dials Warehouse, 44 Earlham Street, London.
Cooper, David (1993), 'Human Sentiment and the Future of Wildlife', *Environmental Values*, Vol. 2, No. 4, pp. 335–346.
Coppinger, Raymond and Coppinger, Lorna (2001), *Dogs: A Startling New Understanding of Canine Origin, Behavior and Evolution*, Scribner, New York.
Coppinger, R. and Schneider, R. (1995), 'Evolution of working dogs' in James Serpell, (ed.) *The Domestic Dog, its Evolution, Behaviour and Interactions with People*, Cambridge University Press, Cambridge.
Darwin, Charles (1985; 1859), *The Origin of Species*, Penguin, Bungay, Suffolk.
Darwin, Charles (1998; 1905), *Variations of Animals and Plants,* Vol. 1, The Johns Hopkins UP, London.
Dawkins, R. (1986), *The Blind Watchmaker*, Penguin, London.
Deakin, R. (1997), 'Questions', *BBC Wildlife*, Vol. 15, No. 4.
De Blieu, Jan (1993), *Meant to be Wild; The Struggle to Save Endangered Species Through Captive Breeding*, Fulcrum Publishing, Colorado.

Defra,http://www.defra.gov.uk/wildlife-countryside/resprog/findings/ruddy/trial.htm.

De Grazia, David (1996), *Taking Animals Seriously; Mental Life and Moral Status*. Cambridge University Press, Cambridge.

Department of the Environment Code of Practice for the Release of Captive Bred Barn Owls in Britain, Section 5.2, p. 10.

Des Jardins, Joseph R. (1997), *Environmental Ethics; An Introduction to Environmental Philosophy*, Wadsworth Publishing Company, ITP, London.

De Waal, Frans (1996), *Good Natured; The Origins of Right and Wrong in Humans and Other Animals*, Harvard University Press, Cambridge, Massachusetts.

Diamond, J. (1984), 'Historical Extinction: A Rosetta Stone for Understanding Prehistoric Extinctions' in P. S. Martin and R. G. Klein (eds), *Quarterly Extinctions: A Prehistoric Revolution*, University of Arizona Press, Tuscon, Arizona.

Dower, John, (1991), *Lifesense, Our Lives Through Animal Eyes*, BBC Books, London.

Dyer, James (1990), *Ancient Britain*, BT Batsford Ltd, London.

Ehrlich and Ehrlich, (1981), *Extinction: The Causes and Consequences of the Disappearance of Species*, Random House, New York.

Elliot, R. (1994), 'Extinction, Restoration, Naturalness', *Environmental Ethics*, 16, pp. 135–143.

Estes, J. A. (1992), 'Catastrophes and Conservation: Lessons from Sea Otters and the Exxon Valdez', *Science*, 254, p. 1596.

Evans, Paul (1996), 'An alien response', *The Guardian*, August 7.

Evernden, P. (1978), 'Beyond Ecology: Self, Place and the Pathetic Fallacy' in *The North American Review*, Winter.

Evernden, P. (1985), *The Natural Alien: Humankind and Environment*, University of Toronto Press, Toronto.

Flew, A. G. N. (1967), *Evolutionary Ethics*, MacMillan, London.

Flowers, Mark (1995), 'Flower Power', *BBC Wildlife*, Vol. 2, No. 1.

Fry, Mike (1996), 'Infant insectivorous bat rehabilitation' in M. Brown (ed), *Bat Care News*, No. 14, June.

Gaston, Kevin (1994), *Rarity*, Chapman & Hall, London.

Ginsburg, Joshua and Cole, Mary (1994), 'Wild at heart', *New Scientist*, November, 19.

Goodall, Jane (1986), *The Chimpanzees of Gombe*, Harvard University Press, Cambridge, Massachusetts.

Gould, S. J. (1977), *Ever Since Darwin*, Penguin, London.

Gould, S. J. (1989), *Wonderful Life*, Penguin, London.

Green, P. (1997), 'How Not to Kill a Deer', *BBC Wildlife*, Vol. 15, No. 5.

Gritten, Rod (1998), 'Alien Invasion (Against)' *BBC Wildlife*, Vol. 16, No. 9.

Groombridge, B. (ed) (1992), *Global Biodiversity: Status of the Earth's Living Resources* (A Report compiled by the World Conservation Monitoring Centre), Chapman and Hall, London.

Grumbine, Edward R. (1994), 'Wildness, Wise Use, and Sustainable Development', *Environmental Ethics*, Vol. 16, Fall, pp. 227–249.

Gunn, Alistair (1980), 'Why Should We Care About Rare Species?', *Environmental Ethics*, Vol. 2, Spring.

Haldane, J. B. S. (1932), *The Causes of Evolution*, Longmans, Green & Co., London.

Harris, M. P. and Wanless, S. (1997), *Sula*, 11, pp. 183–185.

Hearne,Vicki (1987), *Adam's Task: Calling Animals by Name*, Harper Perennial, New York.

HMSO (1994), *Biodiversity: The UK Action Plan*, HMSO, London.

Holland, Alan (1995), 'Philosophising the Green' The Greening of Higher Education Conference (GHECO), 11–13 September, London.

Holland, Alan and O'Neill, John (1996), 'The Integrity Of Nature Over Time', *The Thingmount Working Paper Series on The Philosophy of Conservation*, Department of Philosophy, Lancaster University, Lancaster.

Hollander (1976), 'Discussion No. 5' in Boitani, L (ed.) *Proceedings of the Seminar, Reintroductions: Techniques and Ethics*, World Wildlife Fund, Rome.

Hospers, J. (1970), *An Introduction to Philosophical Analysis*, 3rd Edition, Routledge, London.

Ingold, Tim (1995), 'Life Beyond the Edge of Nature? or, The Mirage of Society', Presented at the University of Lancaster, 1996, in Greenwood, John, D. (ed.) (1996), *The Mark of the Social*, Rowan and Littlefield.

International Otter Survival Fund, 'Welcome to the International Otter Survival Fund' (leaflet), Broadford, Isle of Skye, Scotland.

International Union for Conservation of Nature and Natural Resources (1988), *1988 IUCN Red List of Threatened Animals*, IUCN, Cambridge.

Jersey Wildlife Preservation Trust (1991), 'The Wildlife Preservation Trust' (booklet), Les Augres Manor, Trinity, Jersey, Channel Islands.

Johnson, William (1990), *International Whale Bulletin*, Whale & Dolphin Conservation Society, Summer, No. 6.

Johnson, William (1993), 'Hermit of the Aegean', *BBC Wildlife*, Vol. 11, No. 7.

Jordan, W. (1991), *Divorce Among the Gulls; An Uncommon Look at Human Nature*, Abacus, London.

Katz, Eric (1992), 'The Call of the Wild: The Struggle against Domination and the Technological Fix of Nature', *Environmental Ethics*, Vol. 14, Fall, pp. 265–273.

Kelbie, Paul (2002), 'International rescue mission to evacuate doomed hedgehogs', *The Independent*, December 20.

Kiley-Worthington, Marthe (1997), Philosophy Department Seminar, February 1, Lancaster University.

Kirkwood, J. (1993), 'Intervention for wildlife health, conservation and welfare', *The Veterinary Record*, 132, March 6, pp. 235–238.

Kirkwood, J. K. and Sainsbury A. W. (1996), 'Ethics of Interventions for Welfare of Free-Living Wild Animals', *Animal Welfare*, 5, pp. 235–243.

Lacy, R. (1995), 'Culling Surplus Animals for Population Management', in Norton, B. (ed.), *Ethics on the Ark; Zoos, Animal Welfare, and Conservation*, Smithsonian Institute Press, London.

Lavigne, David (1994), 'Killed, acquitted, captured', *BBC Wildlife*, Vol. 12, No. 9.

Lawson, Trevor (1998), 'Taking Issue: Alien Invasion', *BBC Wildlife*, Vol. 16, No. 9.

Leahy, M. (1996), The Natural History Programme, May 19, *BBC Radio 4*.

Lemon, Tess (1993), 'A chimp in a cage is nothing', *BBC Wildlife*, Vol. 11, No. 3.

Leopold, Aldo (1966; 1949), *A Sand County Almanac; with Essays on Conservation from Round River*, Ballantine Books, New York.

Lieberman, David A. (1993), *Learning; Behaviour and Cognition*, Brooks/Cole, California.

Linklater, Magnus (1997), 'RSPB takes risk on capercaillie's survival', *The Times*, May 10.

Linzey, Andrew (1996), 'Taking Issue: Study and be damned? (Against)', *BBC Wildlife*, Vol. 14, No. 1.

Livingston, John A. (1994), *Rogue Primate: An Exploration of Human Domestication*, Key Porter, Toronto.

Mabey, Richard (1995a), 'Letters', *BBC Wildlife*, Vol. 13, No. 10.

Mabey, Richard (1995b), 'Richard Mabey', *BBC Wildlife*, Vol. 13, No. 11.

Mabey, Richard (1996), 'Richard Mabey', *BBC Wildlife*, Vol. 14, No. 9.

Macdonald, David (1989), *Running with the Fox*, Unwin Hyman, London.

Manning, A. and Dawkins, M. S. (1992), *An Introduction to Animal Behaviour* (4th edition), Cambridge University Press, Cambridge.

Mason, G. and Mendle, M. (1993), 'Why Is There No Simple Way of Measuring Animal Welfare?', *Animal Welfare*, 2, pp. 301–319.

Massei, Giovanna, Lyon, Alicia J. and Cowan, David P. (2002), 'Conditioned taste aversion can reduce egg predation by rats', *Journal of Wildlife Management*, 66 (4), pp. 1134–1140.

Masson, J. (1997), *Dogs Never Lie about Love; Reflections on the Emotional World of Dogs*, Jonathan Cape, London.

Mayer, S. (1996), *A Review of Live Strandings of Cetaceans: Implications for their Veterinary Care, Rescue and Rehabilitation in the UK*, Whale and Dolphin Conservation Society.

Maynard, Robin (1994), 'The penguin predicament', *BBC Wildlife*, Vol. 12, No. 9.

McCune, Sandra, McPherson, Justine A. and Bradshaw, John W. S. (1995), 'Avoiding Problems: The Importance of Socialization', in Robinson, I. (ed.) *The Waltham Book of Human-Animal Interaction: Benefits and Responsibilities of Pet Ownership*, pp. 71–86, Pergamon, Oxford.

McGreal, S. (1993), 'An orphan's choice', *BBC Wildlife*, Vol. 11, No. 3.

Mead, Chris (1997), 'Poor prospects for oiled birds' *Nature*, Vol. 390, December 4.

Mill, John Stuart (1904), *Nature; The utility of religion; and Theism*, Watts & Co, London.

Miller, Lowell A., Johns, Brad E. and Elias, Donald J. (1998), 'Immunocontraception as a wildlife management tool: some perspectives', *Wildlife Society Bulletin*, 26 (2) pp. 237–243.

Morant, P. D., Cooper, J. and Randall, R. M. (1981), 'The rehabilitation of oiled Jackass Penguins *Spheniscus demersus*, 1970–1980' in Cooper, J. (ed.) Proceedings of the Symposium *Birds of the Sea and Shore*, pp. 267–285, African Group, Cape Town, South Africa.

Morgan, Kenneth O. (ed.) (1984), *The Oxford Illustrated History of Britain*, Oxford University Press, Oxford.

Morris, P. A., Meakin, K. and Sharafi, S. (1993), 'The behaviour and survival of rehabilitated hedgehogs (Erinaceus Europaeus)', *Animal Welfare*, 2, pp. 53–66.

Morris, P. A. and Warwick, H. (1994), 'A study of rehabilitated juvenile hedgehogs after release into the wild', *Animal Welfare*, 3, pp. 163–177.
Nantsou, T. and Antipas, M. (1994), 'Sea Turtle Rescue Network in Greece', Sea Turtle Protection Society of Greece, Athens.
Neale and Dean, (1998), *Sunday Telegraph*, March 15.
Nicholson, Adam (1995), 'Moo over darling', *The Sunday Telegraph Review*, December 3.
Northwest Tonight, (1992), BBC 1 Television, February 11.
Norton, B. (1986), 'On the inherent Danger of Undervaluing Species' and 'Epilogue', in Norton B., *The Preservation of Species; The Value of Biological Diversity*, Princeton University Pres, New Jersey.
Norton, Bryan G. (1987), *Why Preserve Natural Variety?* Princeton University Press, New Jersey.
Ollason, J. G. (1991), 'What Is This Stuff Called Fitness?', *Biology and Philosophy*, 6, pp. 81–92.
O'Neill, John (1995), *Ecology, Policy and Politics*, Routledge, London.
Ordish, George and Binder, Pearl (1967), *People and Pigeons*, Dennis Dobson, London.
Peel, John (1995), *BBC Radio 4*, October 7.
Peretti, Jonah H. (1998), 'Nativism and Nature. Rethinking Biological Invasion', *Environmental Values*, 7, pp. 183–192.
Phillips, Dave (1996), 'The Natural History Programme', *BBC Radio 4*, May 19.
Ponting, C. (1992), *A Green History of the World*, Penguin, London.
Porrit (1996), 'Jonathon Porritt' *BBC Wildlife*, 14, No. 1.
Prappas *et al* (1991), *Environmental Biology of Fishes*, 31, (1), pp. 33–40.
Price, E. O. (1984), 'Behavioural Aspects of Animal Domestication' *The Quarterly Review of Biology*, 59, No. 1, pp. 1–32.
Primack, Richard B. (1993), *Essentials of Conservation Biology*, Sinauer Associates Inc., Massachusetts.
Pryor, Karen (1995), *On Behaviour; Essays and Research*, Sunshine Books, Washington.
Rackham, Oliver (1993; 1986), *The History of the Countryside: The Classic History of Britain's Landscape, Flora and Fauna*, J M Dent, London.
Ramsden, David and Frances, 'Barn Owl Reintroduction Report 1986–1988', The Barn Owl Trust, Waterleat, Ashburton, Devon.
Ratcliffe, Jane (1983), *Through the Badger Gate*, Dalesman Books, Clapham, via Lancaster.
Reader's Digest Field Guide to the Animals of Britain. (1984) Reader's Digest Association Limited, London.
Reed, Edward S. (1994), 'The affordances of the animate environment: social science from the ecological point of view' in T. Ingold, (ed.) *What is an Animal?*, Routledge, London.
Regan, Tom (1984), *The Case for Animal Rights*, Routledge, London.
Regan, Tom and Singer, Peter (1989), *Animal Rights and Human Obligations*, Prentice Hall, New Jersey.
Ridley, Matt (1994), 'What on Earth', *BBC Wildlife*, Vol. 12, No. 10.

Robertson, Charles and Harris, Stephen (1991), (Unpublished) 'The Rehabilitation of Orphaned Fox Cubs Summary Report', Grant W – 7/91.

Routh, A. (1995), Wildlife Conference, Lower Moss Wood, Cheshire, August 5.

Routh, A. (1996), 'First aid and rehabilitation of British Bats', *The Rehabilitator*, No. 23, Summer.

RSPB (1991), *Action for Birds and the Environment; Environmental Manifesto for the 1990s*, RSPB, The Lodge, Sandy, Bedfordshire.

Ruse, M. (1982), *Darwinism Defended: A Guide to the Evolution Controversies*, Addison-Wesley, London.

Russell, David (1997), Joint British Association for Nature Conservation and Lancaster University (BANC/LANC) Conference, Ambleside.

Russell, David (1998), Conference, Lancaster House Hotel, Lancaster University, March 10.

Schmidt, K. (1997), 'A drop in the ocean', *New Scientist*, May 3, pp. 41–44.

Scott, Michael (1989), 'Unhappy homecomings', *BBC Wildlife*, Vol. 7, No. 12.

Scott, Michael (1991), 'Cull to be kind', *BBC Wildlife*, Vol. 9, No. 12.

Seboek, Thomas, A. (1994), '"Animal" in biological and semiotic perspective' in T. Ingold, (ed) *What is an Animal*? Routledge, London.

Sgeulaich, L. (1993), 'House Martens', *BBC Wildlife*, Vol. 11, No. 9.

Sharp, B. (1996), 'Post-release survival of oiled, cleaned seabirds in North America', *IBIS*, 138, pp. 222–228.

Shimony, A. (1989), 'The Non-Existence Of The Principle Of Natural Selection' *Biology and Philosophy*, 4.

Shrader-Frechette, K. S. and McCoy, E. D. (1994), 'Biodiversity, Biological Uncertainty, and Setting Conservation Priorities', *Biology and Philosophy*, 9.

Singer, Peter (1975), *Animal Liberation; Towards an End to Man's Inhumanity to Animals*, Thorsons Publishers Limited, Northamptonshire.

Sober, E. (1984), *The Nature of Selection: Evolution Theory in Philosophy*, MIT Press, Cambridge, Massachusetts.

Steinhart, Peter (1990), 'Humanity without Biology', *Audobon Magazine*, May.

Stocker, Les (1989), *Something in a Cardboard Box; Tales from the Wildlife Hospital*, Chatto & Windus, London.

Summers-Smith, Denis (1997), 'Sparrows rule, OK?' *BBC Wildlife*, Vol. 15, No. 2.

Sylvan, R. 'Mucking with Nature', unpublished paper.

Tickell, Oliver (1995), 'Return to the wild', *Geographical*, February.

Tudge, C. (1991), *Last Animals at the Zoo; How Mass Extinction can be Stopped*, Hutchinson Radius, London.

Turner, J. (1994), 'Animals of another earth', *BBC Wildlife*, Vol. 12, No. 4.

Valentine, Wendy (1997), *Caduceus*, Issue 37, Autumn.

Vermeij, G. J. (1986) 'The Biology of Human-Caused Extinction' in B. Norton (ed.) *The Preservation of Species; The Value of Biological Diversity*, Princeton University Press, New Jersey.

Vines, G. (1997), 'Who's suffering now?' *New Scientist*, March 22, pp. 30–33.

Vogel, Steven (1996), *Against Nature: The Concept of Nature in Critical Theory*, State University of New York Press, New York.

Warburton, Tony (1995), Wildlife Conference, Lower Moss Wood, Cheshire, August 5.

Warwick, Hugh (1993), 'Confessions of a hedgehogger' *BBC Wildlife*, Vol. 11, No. 7.

Waters, C. K. (1986), 'Natural Selection Without Survival Of The Fittest' *Biology and Philosophy*, 1, pp. 207–225.

Watson, D. (1989), 'Red Deer in Scotland – A Resource out of Control', *ECOS*, 10.

Western, Sue (1994), 'Ape Opera', *BBC Wildlife*, Vol. 12, No. 1.

Whitcher, Derek (1995), Wildlife Conference, Lower Moss Wood, Cheshire, August 5.

Wigan, Michael (1993), *Stag at Bay; The Scottish Red Deer Crisis*, Swan Hill Press, Shrewsbury.

Wilcox, B. (1988), 'Tropical Deforestation and Extinction', in International Union for Conservation of Nature and Natural Resources, *IUCN Red List of Threatened Animals*, IUCN, Cambridge.

Williams, Martin (1992), 'A new start in California', *BBC Wildlife*, Vol. 10, No. 3.

Wilson, Bradley E. (1996), 'Changing conceptions of species', *Biology and Philosophy*, 11, pp. 405–420.

Wilson, Edward O. (1992), *The Diversity of Life*, Penguin, London.

Wittgenstein, Ludwig (1997), *Philosophical Investigations*, Blackwell, Oxford.

Woolpy, J. H. and Ginsburg, B. E. (1967), 'Wolf socialization; A study of temperament in a wild social species', *American Zoologist*, 7, pp. 357–363.

Worster, Donald (1985), *Nature's Economy: A History of Ecological Ideas*, Cambridge University Press, Cambridge.

Yalden, D. W. (1982), 'When did the mammal fauna of the British Isles arrive?', *Mammal Review*, Vol. 12, No. 1, March.

Zimen, E. (1987), 'Ontogeny of approach and flight behavior towards humans in wolves, poodles and wolf-poodle hybrids', in H. Frank, (ed.) *Man and Wolf*, Dr. W. Junk Publishers, pp. 275–292, Dordrecht, the Netherlands.

Index

For Product Safety Concerns and Information please contact our EU
representative GPSR@taylorandfrancis.com
Taylor & Francis Verlag GmbH, Kaufingerstraße 24, 80331 München, Germany

www.ingramcontent.com/pod-product-compliance
Ingram Content Group UK Ltd.
Pitfield, Milton Keynes, MK11 3LW, UK
UKHW022034061126
477531UK00008B/321
9781138618701